Structural Engineering for Architects: A Handbook

Pete Silver
Will McLean
Peter Evans

Laurence King Publishing

Structural Engineering for Architects: A Handbook

Published in 2013
by Laurence King Publishing Ltd
361–373 City Road
London EC1V 1LR
Tel +44 (0)20 7841 6900
Fax +44 (0)20 7841 6910
E enquiries@laurenceking.com
www.laurenceking.com

Design copyright © 2013
Laurence King Publishing Limited
Text © 2013 Pete Silver, Will
McLean and Peter Evans

Pete Silver, Will McLean and Peter
Evans have asserted their right
under the Copyright, Designs
and Patents Act 1988, to be
identified as the Authors of this
work.

All rights reserved. No part of this
publication may be reproduced
or transmitted in any form or by
any means, electronic or
mechanical, including photocopy,
recording or any information
storage and retrieval system,
without prior permission in
writing from the publisher.

A catalogue record for this book is
available from the British Library

ISBN 978 178067 055 3

Designed by Hamish Muir

Printed in China

UNIVERSITY
OF SHEFFIELD
LIBRARY

Contents

Introduction

'At the age of 17 I was told that I could never be an architect, as I would never fully comprehend building structures. So that is how I came to study architecture, with a chip on my shoulder. I religiously attended all the lectures on structural engineering, indeed any engineering, and found out that it was surprisingly easy to understand and, even better, that it was fun. Subsequently I fell in love with the engineering science, not that I have ever fully comprehended it but – who cares? You don't need to "understand" love after all....

This book is one of those love letters that one receives and only has to decide if one wants to respond. How much I wish I had come across this book in my youth – it would have saved so much effort spent reading so many boring ones.

You can take it or leave it, but since it is now available no-one can now say that "you will never understand structure". Take my word, this book will give another dimension to your understanding of the planet we live on and above all...it's fun!'
Eva Jiricna
June 2011

The aim of this book is to enable students of architecture to develop an intuitive understanding of structural engineering so that, in the long term, they are able to conduct productive dialogues with structural engineers. It is also hoped that the book will serve as a valuable reference and sourcebook for both architecture and engineering.

In Giorgio Boaga's book *The Concrete Architecture of Riccardo Morandi*, published in 1965, the Italian engineer Morandi discusses the perceived difficulty of the architect–engineer relationship, but refuses to take sides in this unhelpful argument. More importantly, he describes how '...it is always possible, within certain limits, to solve a problem – functionally, structurally and economically – in several equally valid ways' and that '...the loving care given to the formal details (quite independently of the requirements of calculation) transcend the purely technical aspect and, intentionally or not, contribute to artistic creation.'[1] In these statements Morandi is no more siding with the gifted 'calculator' than with the flamboyant designer – he is merely on the side of interesting work, which may appear unnervingly simple or unexpectedly expressive.

In his 1956 book *Structures*, Pier Luigi Nervi explains his use of isostatic ribs, which followed the stress patterns that had been made visible by new photoelastic imagery techniques. More recently, the detailed arithmetic and algebraic calculations of Finite Element Analysis (FEA) have been made visible through computer graphical output – an incredibly powerful tool for the more intuitively minded. A step further than this is structural engineer Timothy Lucas's putative explorations of a digital physical feedback system, which would enable the engineer to physically differentiate and explore the structural forces through an augmented physical model. Throughout the history of technology, physical testing has been and continues to be a vital component in the development of technology and design engineering strategies. Similarly, the field of biomimetics is surely only an academic formalization of a timeless process, where we learn from the rapid prototyping of nature and the previous unclaimed or forgotten inventions of man to develop new design, engineering, material and operational strategies.

The book is divided into four parts:

Part 1 – Structures in nature describes some common structural forms found in nature.

Part 2 – Theory outlines a general theory of structures and structural systems that are commonly applied to the built environment.

Part 3 – Structural prototypes introduces methods for developing and testing structural forms, including both 'hands-on' modelmaking and full-scale prototypes, as well as analytical computer modelling.

Part 4 – Case studies presents a selection of key figures involved in the evolution of structural engineering and built form, from the mid-nineteenth century to the present.

1 Boaga, G., and Boni, B., *The Concrete Architecture of Riccardo Morandi*, London: Alex Tiranti, 1965, p. 10

1
Structures in nature

1.1 Tree

More than 80,000 species of tree, ranging from arctic willows a few centimetres high to giant redwoods that can grow to over 100 metres tall, cover 30 per cent of the Earth's dry land.

Structure

Trees come in various shapes and sizes, but all possess the same basic structure. They have a central column, the trunk, which supports a framework of limbs, branches and twigs. This framework is called the crown, and it is estimated that there are a finite number of branching systems for all tree species (around 30). Branches and twigs in turn have an outside covering layer of leaves. A tree is anchored in the ground using a network of roots, which spread and grow thicker in proportion to the growth of the tree above the ground.

All parts of the framework of a tree – trunk, branches and twigs – are structural cantilevers with flexible connections at the junctions. All have the property of elastic behaviour.

Hardwood and softwood: these terms refer to the types of tree from which the wood comes. Hardwood comes from deciduous forests; softwood from coniferous forests. Although hardwoods are generally of a higher density and hardness than softwoods, some (e.g. balsa) are softer.

Growth

Much of the energy produced by the leaves of a tree has to be diverted to make unproductive tissue (such as the woody trunk, branches and roots) as the tree grows. The overwhelming portion of all trees (up to 99 per cent) is made up of non-living tissue, and all growth of new tissue takes place at only a few points on the tree: just inside the bark and at the tips of the twigs and roots. Between the outer (cambial) layer and the bark there is an ongoing process of creating sieve tubes, which transport food from the leaves to the roots. All wood is formed by the inner cambium and all food-conveying cells are formed by the outer cambium.

A tree trunk grows by adding a layer of new wood in the cambium every year. Each layer of new wood added to a tree forms a visible ring that varies in structure according to the seasons. A ring composed of a light part (spring growth) and a dark part (late summer/autumn growth) represents one year's growth. Timber used in construction is chosen on the basis of having an even balance of stresses within the plank. If a tree has grown on the side of a hill, it will grow stronger on one side and the stresses will be locked in to create a harder 'red' wood that will eventually cause a plank to warp – by twisting or bowing.

Wind resistance

Trees are generally able to withstand high winds through their ability to bend, though some species are more resilient than others. Wind energy is absorbed gradually, starting with the rapid oscillation of the twigs, followed by the slower movement of the branches and finally through the gently swaying limbs and trunk. The greater surface area of a tree in leaf makes it more susceptible to failing under wind load.

1

1
The basic structure
of a tree

2
Section through a tree trunk
a outer bark
b inner bark
c heartwood
d cambium
e sapwood

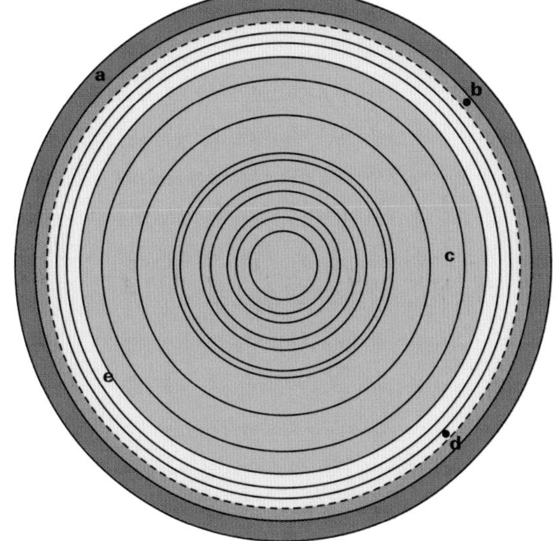

2

1.2
Spider's web

Material properties

Spider silk is also known as gossamer and is composed of complex protein molecules. Chains of these molecules, with varying properties, are woven together to form a material that has an enormous capacity for absorbing energy. The silk of the Nephila spider is the strongest natural fibre known to man.

A general trend in spider-silk structure is a sequence of amino acids that self-assemble into a (beta) sheet conformation. These sheets stack to form crystals, whereas the other parts of the structure form amorphous areas. It is the interplay between the hard crystalline segments and the elastic amorphous regions that gives spider silk its extraordinary properties. This high toughness is due to the breaking of hydrogen bonds in these regions. The tensile strength of spider silk is greater than the same weight of steel; the thread of the orb-web spider can be stretched 30–40 per cent before it breaks.

Silk production

Spiders produce silken thread using glands located at the tip of their abdomen. They use different gland types to produce different silks; some spiders are capable of producing up to eight different silks during their lifetime.

Web design and production

Spiders span gaps between objects by letting out a fine adhesive thread to drift on the breeze across a gap. When it sticks to a suitable surface at the far end, the spider will carefully walk along it and strengthen it with a second thread. This process is repeated until the thread is strong enough to support the rest of the web. The spider will then make Y-shaped netting by adding more radials, while making sure that the distance between each radial is small enough to cross. This means that the number of radials in a web is related directly to the size of the spider and the overall size of the web. Working from the inside out, the spider will then produce a temporary spiral of non-sticky, widely spaced threads to enable it to move around its own web during construction. Then, beginning from the outside in, the spider will replace this spiral with another, more closely spaced one of adhesive threads.

Impact resistance

The properties of spider silk allow it to be strong in tension, but also permit elastic deformation. When completed, the entire spider web is under tension; however, the elastic nature of the fibres enables it to absorb the impact of a fast-flying insect. On impact a local oscillation will occur, and the faster the oscillation the greater the resistance. This ability to store energy, and the fact that most of the energy is dissipated as the fibre deforms, allows spiders to intercept and catch their prey, by absorbing these creatures' kinetic energy.

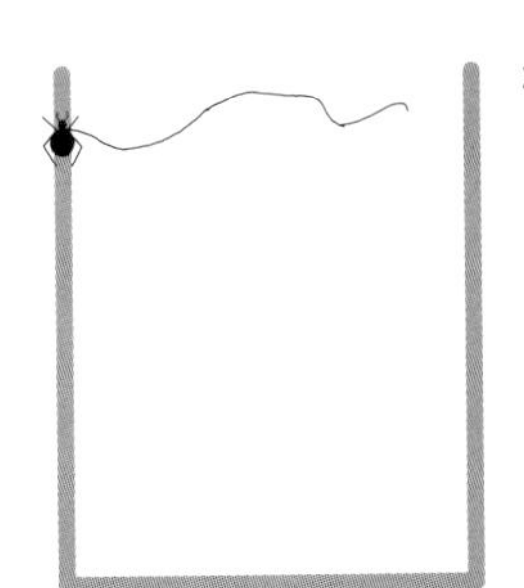

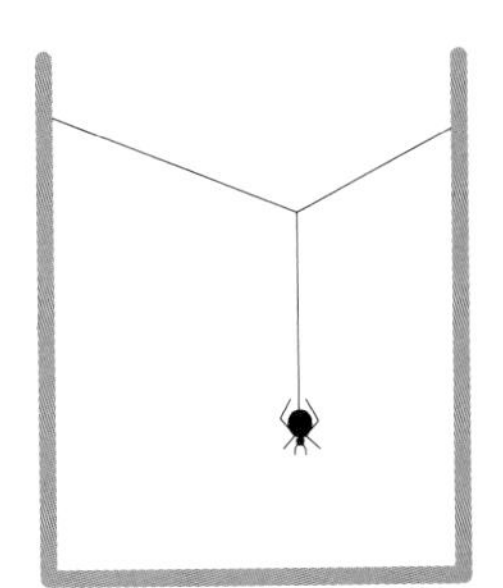

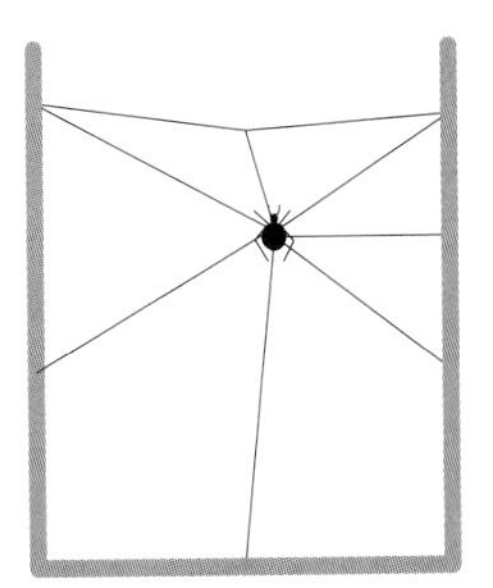

1
The spider's silk-spinning glands

2
Sequence of web building

3
A giant spider web

4
The successful completion of an arrested landing on the flight deck of an aircraft carrier. The 'checkmates' to which the aircraft becomes attached perform a similar kind of impact resistance to that of a spider's web.

1.3 Eggshell

The structure of an eggshell varies widely among species but it is essentially a matrix lined with mineral crystals, usually a compound such as calcium carbonate. It is not made of cells, and harder eggs are more mineralized than softer ones.

Bird's eggs – material properties

Birds are known for their hard-shelled eggs. The eggshell comprises approximately 95 per cent calcium carbonate crystals, which are stabilized by an organic (protein) matrix. Without the protein, the crystal structure would be too brittle to keep its form.

Shell thickness is the main factor that determines strength. The organic matrix has calcium-binding properties and its organization during shell formation influences the strength of the shell: its material must be deposited so that the size and organization of the crystalline (calcium carbonate) components are ideal, thus leading to a strong shell. The majority of the shell is composed of long columns of calcium carbonate.

The standard bird eggshell is a porous structure, covered on its outer surface with a cuticle (called the bloom on a chicken egg), which helps the egg retain its water and keep out bacteria.

In an average laying hen, the process of shell formation takes around 20 hours.

Strength and shape

The structure of a bird's eggshell is strong in compression and weak in tension. As weight is placed on top of it, the outer portion of the shell will be subject to compression, while the inner wall will experience tension. The shell will thus resist the load of the mother hen. Young chicks are not strong, but by exerting point-load forces on the inside of the shell they are able to break out unaided (the chick has an egg-tooth, which it uses to start a hole).

It is the arch/dome shape of the eggshell that helps it resist tension.

The strength of the dome structure of an eggshell is dependent on its precise geometry – in particular, the radius of curvature. Pointed arches require less tensile reinforcement than a simple, semicircular arch. This means that a highly vaulted dome (low radius of curvature) is stronger than a flatter dome (high radius of curvature). That is why it is easy to break an egg by squeezing it from the sides but not by squeezing it from its ends; staff members at the Ontario Science Centre in Toronto were successful in supporting a 90 kilogram person on an unbroken egg.

1

2

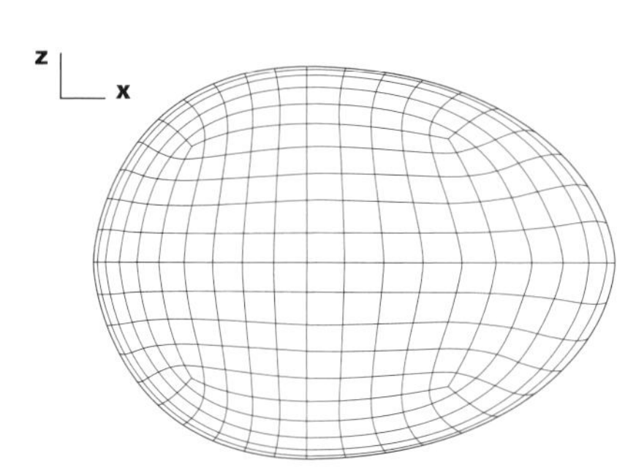

3

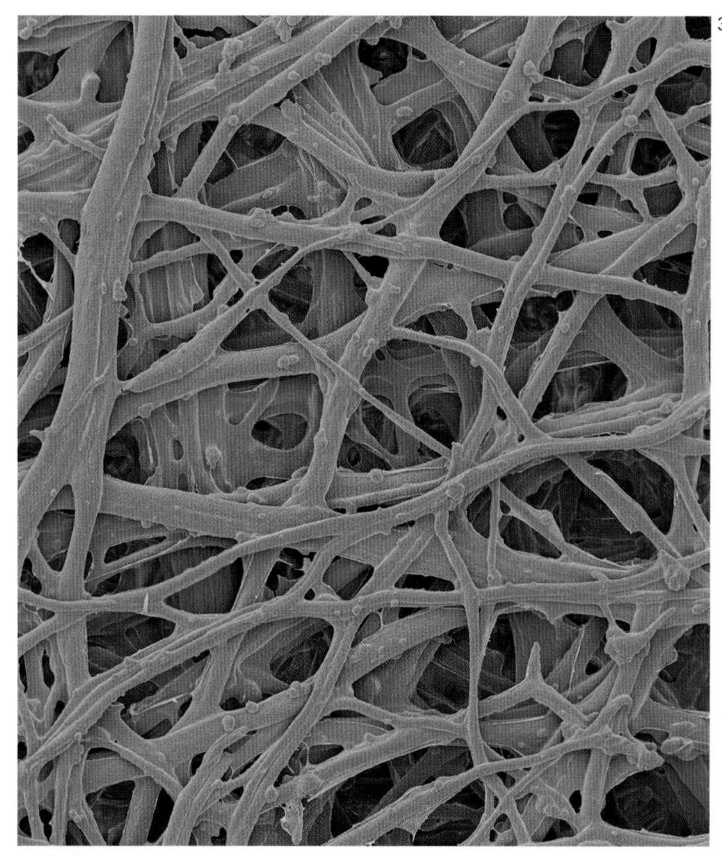

4

5

1
A chicken egg

2
Generated eggshell mesh using shell-type elements

3
A microscopic view of the lattice structure of an eggshell

4
A low-tensile, compressive arch will resist larger forces when pointed

5
The stone and steel arches of the Pavilion of the Future, built by Peter Rice for the 1992 Seville Expo, express their resistance to forces by separating the tensile and compressive elements

1.4
Soap bubbles

Surface tension

A soap bubble exists because the surface layer of a liquid has a certain surface tension that causes the layer to behave elastically. A bubble made with a pure liquid alone, however, is not stable, and a dissolved surfactant such as soap is needed to stabilize it; soap acts to decrease the water's surface tension, which has the effect of stabilizing the bubble (via an action known as the Marangoni effect): as the soap film stretches, the surface concentration of soap decreases, which in turn causes the surface tension to increase. Soap, therefore, selectively strengthens the weakest parts of the bubble and tends to prevent it from stretching further.

Shape

The spherical shape of a soap bubble is also caused by surface tension. The tension causes the bubble to form a sphere, as this form has the smallest possible surface area for a given volume. A soap bubble, owing to the difference in outside and inside pressure, is a surface of constant mean curvature.

Merging

When two soap bubbles merge, they will adopt the shape with the smallest possible surface area. With bubbles of similar size, their common wall will be flat. Smaller bubbles, having a higher internal pressure, will penetrate into larger ones while maintaining their original size.

Where three or more bubbles meet, they organize themselves so that only three bubble walls meet along a line. Since the surface tension is the same in each of the three surfaces, the three angles between them must be equal to 120 degrees. This is the most efficient choice, and is also the reason that cells of a beehive have the same 120-degree angle and form hexagons. Two merged soap bubbles provide the optimum way of enclosing two given volumes of air of different size with the least surface area. This has been termed 'the double bubble theorem'.

1

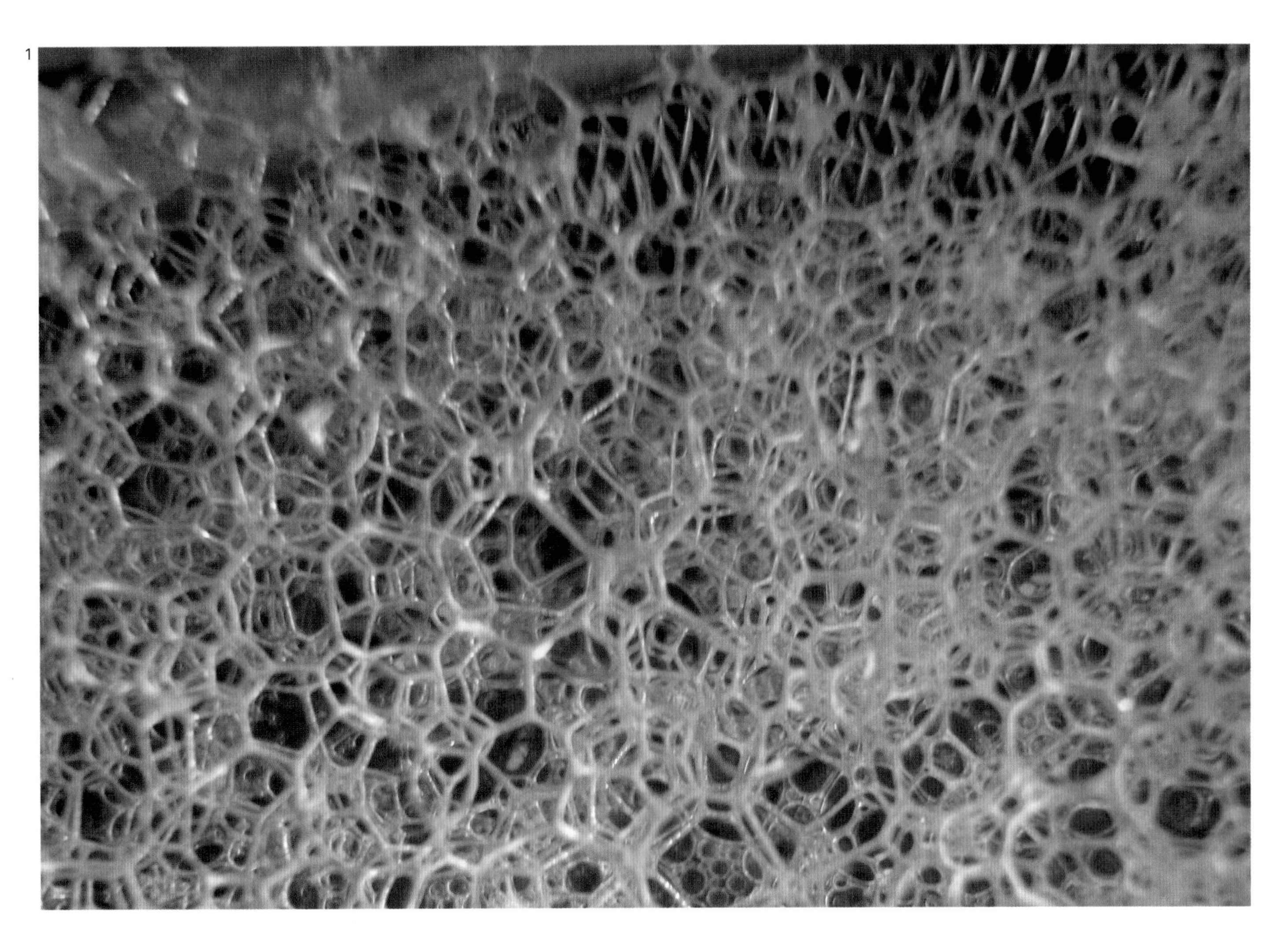

2

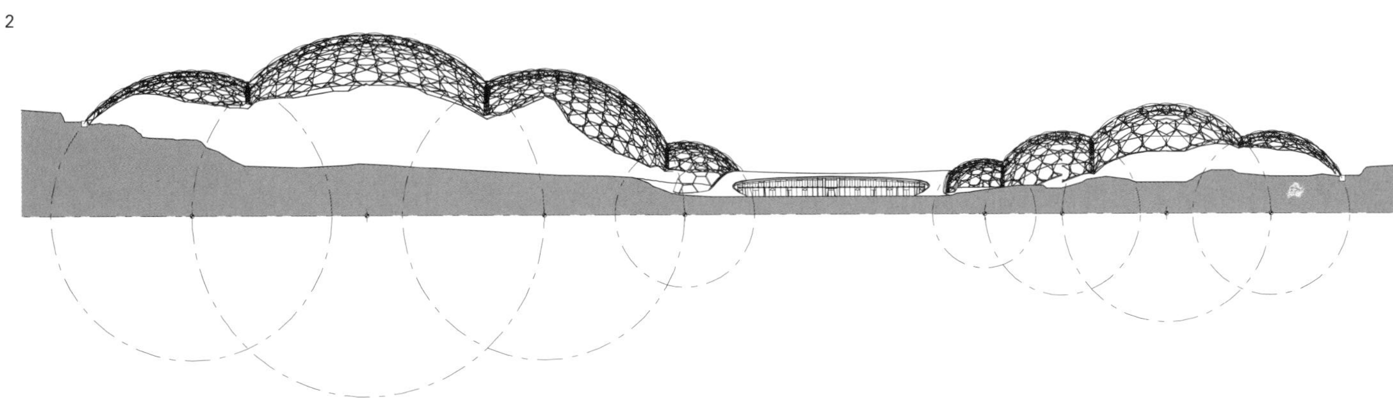

1
Merged soap bubbles

2
The double bubble theorem applied to the design of the bio-domes at the Eden Project in Cornwall, UK, by Nicholas Grimshaw and Partners

1.5
Human body

Human skeleton

The human skeleton has 206 bones that form a rigid framework to which the softer tissues and organs of the body are attached. Vital organs are protected by the skeletal system.

The human skeleton is divided into two distinct parts. The axial skeleton consists of bones that form the axis of the body – neck and backbone (vertebral column) – and support and protect the organs of the head (skull) and trunk (sternum and rib cage). The appendicular skeleton is composed of the bones that make up the shoulders, arms and hands – the upper extremities – and those that make up the pelvis, legs and feet – the lower extremities.

Bones – material properties

Most bones are composed of both dense and spongy tissue. Compact bone is dense and hard, and forms the protective exterior portion of all bones. Spongy bone is found inside the compact bone, and is very porous (full of tiny holes). Bone tissue is composed of several types of cells embedded in a web of inorganic salts (mostly calcium and phosphorus) to give the bone strength, and fibres to give the bone flexibility. The hollow nature of bone structure may be compared with the relatively high resistance to bending of hollow tubes as against that of solid rods.

Muscles – bodily movement

The skeleton not only provides the frame that holds our bodies in shape, it also works in conjunction with the body's 650 muscles to allow movement to occur. Bodily movement is thus carried out by the interaction of the muscular and skeletal systems. Muscles are connected to bones by tendons, and bones are connected to each other by ligaments. Bones meet one another with a joint; for example, the elbow and knee form hinged joints, while the hip is a ball-and-socket type of joint. The vertebrae that go to make the spinal column are connected with an elastic tissue known as cartilage.

Muscles that cause movement of a joint are connected to two different bones, and contract to pull them together. For example, a contraction of the biceps and a relaxation of the triceps produces a bend at the elbow. The contraction of the triceps and relaxation of the biceps produces a straightening of the arm.

Tensegrity

It has been said that the human body, when taken as a whole, is a tensegrity structure. In a tensegrity structure, the compression elements do not touch each other insomuch as they are held in space by separate tension elements (strings, wires or cables). The cell biologist and founding director of the Wyss Institute at Harvard, Don E. Ingber, has made the connection between the tensegrity structures of Kenneth Snelson (see page 156) and living cells, and asserts that 'an astoundingly wide variety of natural systems, including carbon atoms, water molecules, proteins, viruses, cells, tissues and even humans and other living creatures are constructed using a common form of architecture known as tensegrity'.[1]

1 Ingber, Donald, E., 'The Architecture of Life' in *Scientific American*, pp. 48–57, January 1998

1

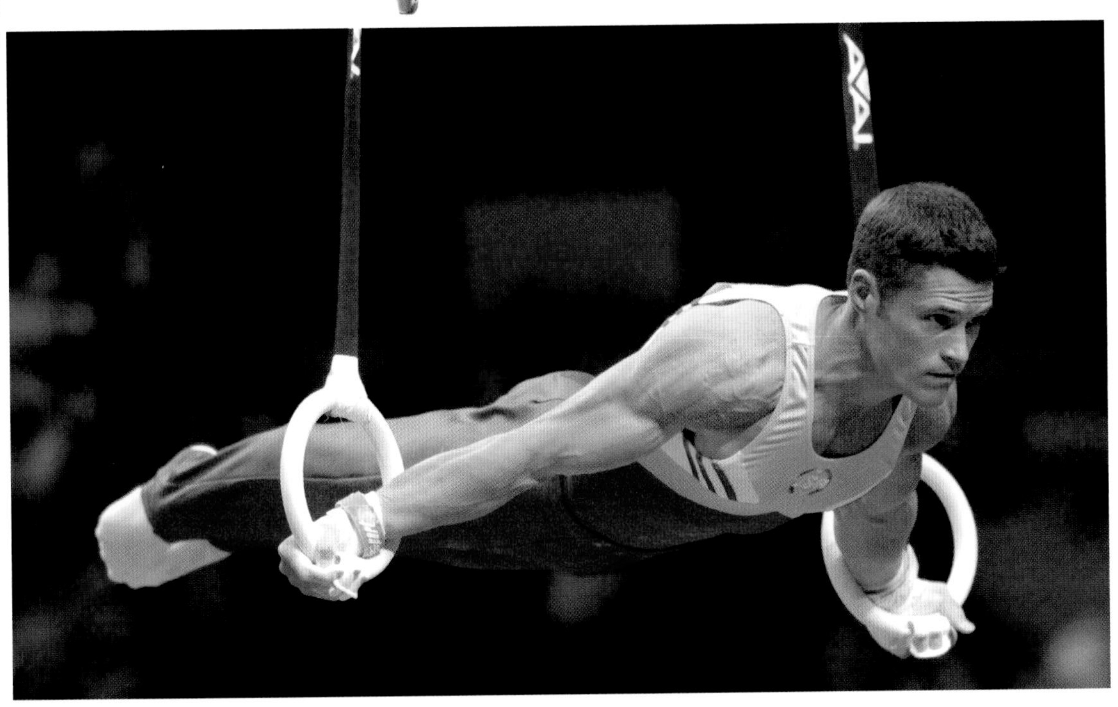

2

1 Ballet pose
Walking is actually 'falling with style'. If you try to walk very slowly, you will start to fall. Try leaning forward from the hips. At some point, your centre of gravity goes 'outside of you', and one leg moves forward to form a triangle that keeps you from toppling over – keeps you stable. Carry on bending, and you will reach the point when the only way to maintain your centre of gravity is to extend your other leg behind you. This is a process known as 'cantilevering'. With built structures, a cantilever describes an element that projects laterally from the vertical. It relies on counterbalance for its stability and on triangulation to resist the bending moments and shear forces of the (canti-) lever arms

2 Gymnastics rings
The stressing of the human body as it strives to maintain a double cantilever

3

3 Tower of people
A Spanish tradition (*torres humanas*), whose intention is self-evident. A number of strategies may be employed, but in all cases a decent foundation for the tower is critical. As with a tree, there is a uniform root structure that is acting to buttress the 'column'. Every participant wears a wide belt to reinforce the connection between the spinal column and the pelvis, and hence protect the kidneys from undue pressure.

4

4 People circle
A circle of people sitting on each others' laps creates a type of tensegrity structure, by which they are all supported without the need for any furniture.

5

5 Flying buttress
The structural principle of the human tower is also expressed in the flying buttresses traditionally used to brace low-tensile masonry structures.

6

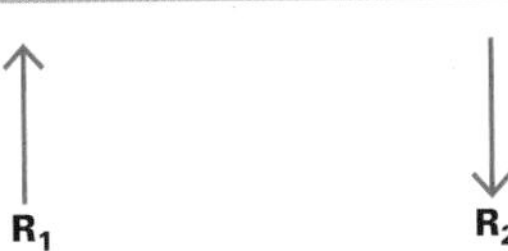

6 Forth Rail Bridge
The designers of the Forth Rail Bridge used their own bodies to demonstrate how the span of the bridge uses the cantilever principle. Replicated here, the bodies of the two men at ground level are acting as columns (in compression), and their arms are being pulled (in tension). The sticks are in compression and are transferring the load back to the chairs.

T = Tension
C = Compression
R = Reaction

2 Theory

2.1
General theory of structures

2.1.1
Introduction

In structural engineering terms, a building can be considered as a series of individual interconnected components whose function is to transfer externally applied loads through a structural system into the building's foundations.

This chapter examines the types of loads that can be applied to structures and the forces that develop within structural components to resist these externally applied loads.

Structural engineering uses the principles of static equilibrium to analyse load distribution. In this chapter the basic concepts of static equilibrium are examined and explained using simple models, while some common mathematical formulae are provided for common beam arrangements.

To determine whether a structural component is capable of resisting the loads applied, to it two major factors have to be considered: the component's size, and the material from which it is made. Further sections of this chapter examine both the geometric and material properties of structural components and their implications on structural performance.

While a building's components must be designed to ensure that they are capable of withstanding the load applied without collapsing, they must also be designed to ensure they can perform their desired purpose without wobbling, deflecting or vibrating to such an extent as to disturb the building's occupants or cause damage to fittings and fixtures. These criteria are often called 'in service' or 'serviceability' states and are explained in the section in this chapter entitled 'Fitness for purpose'.

Individual components are combined to form structures that vary from thin concrete shells to steel-trussed bridges to igloos to multistorey high-rise towers, and all must be sufficiently stable to resist any imposed lateral forces and hence avoid 'falling over'. Stability and the various load-transfer mechanisms different building types employ to achieve stability are explained in this chapter using the building classifications developed by Heinrich Engel.

A brief glossary of the terms used in this section is as follows:

Force – A measure of the interaction between two bodies. Measured in newtons (N) or kilonewtons (kN).

Load – A force acting on a structural element. Measured in newtons (N) or kilonewtons (kN).

Mass – A measure of the amount of material in an object. Measured in kilograms (kg).

Sigma (Σ) – Mathematical term meaning 'the sum of'. For example: $\Sigma F = F_1 + F_2 + F_3$

Weight – A measure of the amount of gravitational force acting on an object. Measured in newtons (N) or kilonewtons (kN) where 1kN = 1000N.

The mass of an object can be converted into weight using the equation;

$$\text{Force} = \text{mass} \times g$$

where g is acceleration due to gravity = $9.81 m/s^2$

Hence a 10kg mass induces a load of:

$$10 \times 9.81 = 98.1 \text{ newtons}$$
(or 0.0981 kilonewtons).

2.1.2 External loads

When external, dead, live or wind loads are applied to a building they induce internal forces within the structural elements that are transmitted into and resisted by the foundations.

Newton's third law of motion states that forces occur in pairs with each force of the pair being equal in magnitude and opposite in direction to the other. Hence, for a building to be stable every external load or force that is applied to it has to be resisted by an equal and opposite force at the supports. This state is called static equilibrium.

External loads can be applied to a structural member in two fundamentally different ways:

Axially – These loads act in the direction parallel to the length of a member and typically induce either internal compressive or tensile forces within it.

Perpendicularly – Perpendicular (or shearing) loads act perpendicularly to the direction of the length of a member. This type of load can induce shear, bending and torsional forces within a member depending on the geometry of the member and point of application of the load.

Each of the five internal forces induced by externally applied loads – tension, compression, shear, bending and torsion – are explained in the following section.

2.1.3
Internal forces

The process of structural analysis and design involves determining the magnitude of the various internal forces (compression, tension, shear, bending and torsion) to which each member is subjected to ensure each member is capable of resisting those forces.

2.1.3.1
Axial

External compressive point load applied to column

External tensile point load applied to a tie member

Axial loads act in the direction parallel to the length of a member. They can either act to resist compressive loads, which try to shorten a member or resist tensile loads, which try to lengthen the member. Members in structural systems that are under compressive loads are termed struts or, if they are vertical, columns. Members under tensile loads are termed ties.

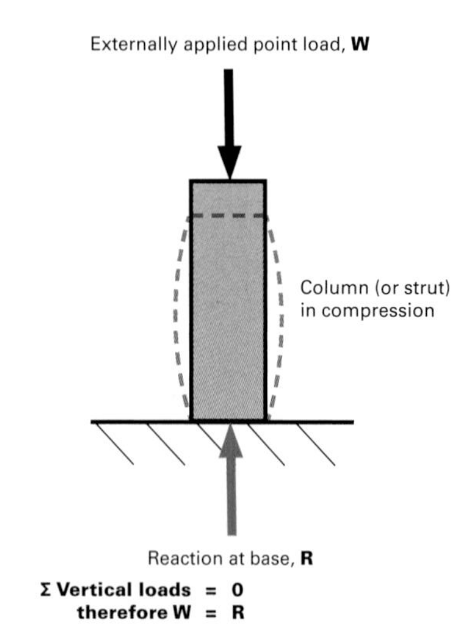

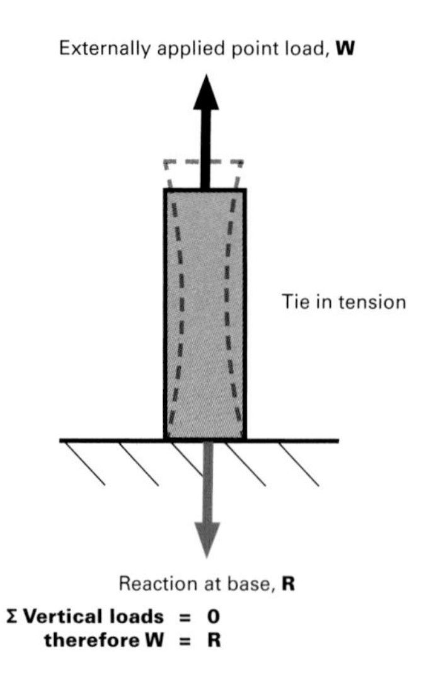

2.1.3.2
Shear

External point load applied to beam

Internal shear forces act perpendicularly to the direction of the length of a member and are induced by externally applied shear loads. For the purposes of analysis, shear loads are considered to be applied to a structural member via either point loads, such as a person standing on a beam, or distributed loads, such as the weight of a floor supported by a beam.

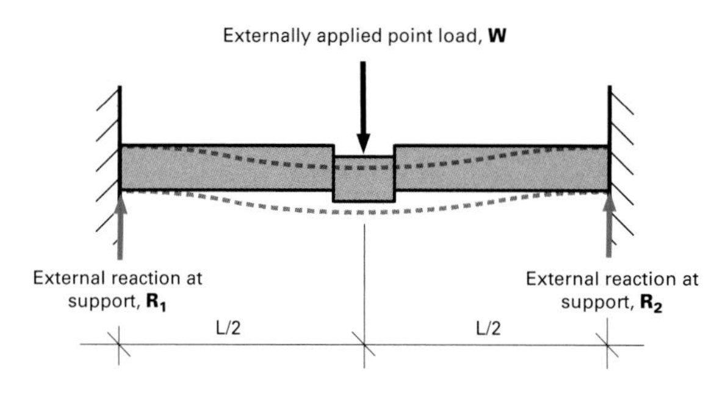

Σ Vertical loads = 0
therefore W = $R_1 + R_2$

2.1.3.3
Bending

External point load applied to beam developing bending moment

Bending, also termed flexure, occurs when a load is applied perpendicularly to the longitudinal axis of a member. This load induces internal forces that act parallel to the length of a member. The magnitude of these internal forces varies proportionally across the depth of the member from compression at one face to tension at the other. At a point between the compression and tension faces the internal force is zero. This is termed the neutral axis. The algebraic sum of the internal forces multiplied by the distance from the neutral axis is called the bending moment. Moments normally occur simultaneously with shear forces and are measured in kilonewton metres (kNm). A simple example of a bending load moment can be demonstrated via a vertical shear load applied to the end of a cantilevering beam. In this situation the bending moment can be calculated as the applied shear load multiplied by the length of the cantilever.

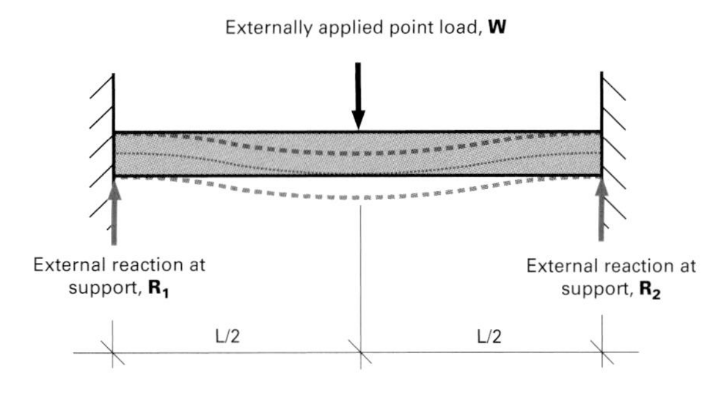

2.1.3.4
Torsion

External point load applied to cantilevering beam developing torsion at support

If the point of application of a load is 'eccentric' from the longitudinal axis of the member, a twisting moment will be developed. This in turn induces torsional forces within the member to resist the twisting action. Torsional forces are distributed across the cross-section of a member in a circular manner where the outer fibres experience the highest forces. The magnitude of torsion is a product of the applied load and distance from the point of application to the longitudinal axis of the member. Torsion is measured in kilonewton metres (kNm).

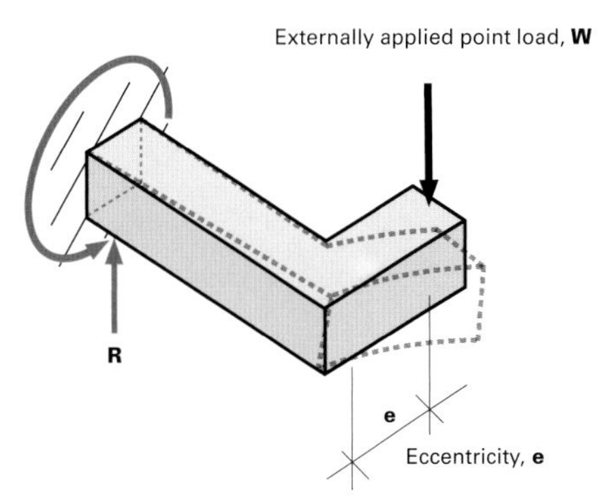

Σ Vertical loads = 0
therefore W = R

Torsion developed at support,
T = W x e

Also, new bending moment developed at support,
M = W x L

2.1.3.5
Static equilibrium

As stated, the applied loads on any structure must be resisted by equal and opposite forces to achieve static equilibrium and thus adhere to Newton's third law of motion. This concept can be demonstrated with a simple seesaw (see illustrations below).

For the seesaw to be in equilibrium both of the following conditions need to be met:

i) The sum of the applied vertical loads are resisted by equal and opposite reaction forces.

Hence $W_1 + W_2 = R$

ii) The sum of the moments around any arbitrary point is zero.

$$\Sigma M = 0$$

For a seesaw to be in static equilibrium the applied vertical forces must be equal to the vertical reaction force:

Hence, $W_1 + W_2 = R$

Also the sum of the applied bending moments around any point must be zero. Hence considering the anti-clockwise bending moment developed around the pivot point;

$$M_{anticlock} = W_1 \times L_1$$

And the clockwise bending moment around the same point:

$$M_{clock} = W_2 \times L_1$$

If the seesaw is balanced i.e. it is in static equilibrium, then:

$$M_{anticlock} = M_{clock}$$

$$W_1 \times L_1 = W_2 \times L_1$$

If these conditions are not achieved the seesaw will 'fail' by falling to the ground. Further examples of balanced systems are ilustrated on the opposite page.

The support reactions to a beam with a single point load can be calculated using the concepts of static equilibrium by considering a beam with a single point load to be an inverted seesaw (i.e. the applied load on the beam is the support reaction to the seesaw and the beam support reactions are the seesaw applied loads). The support reactions generated from a point load of any magnitude placed on the beam can be calculated as indicated on the loaded beam opposite.

The concept of static equilibrium is fundamental to the analysis of structural systems. The following section contains an example of an analysis technique called the 'Method of Sections'. This indicates how the concepts of static equilibrium can be used to calculate the forces in the internal members of a loaded truss.

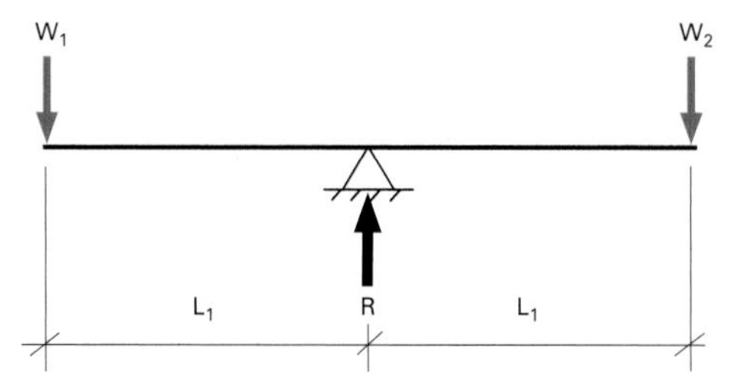

i) System in static equilibrium: $W_1 + W_2 = R$

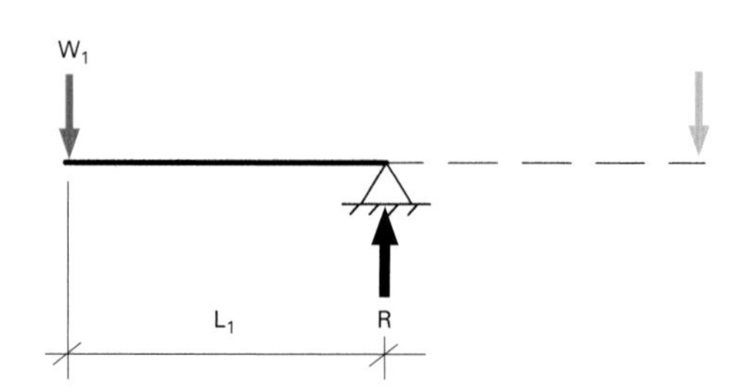

ii) Anticlockwise moment around pivot point,
$M_{anticlock} = W_1 \times L_1$

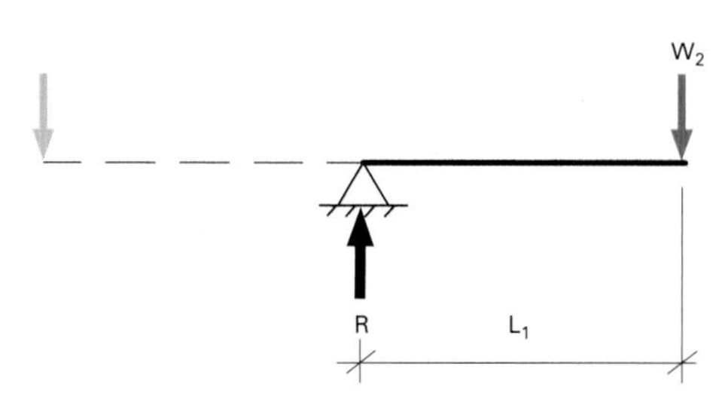

iii) Clockwise moment around pivot point,
$M_{clock} = W_2 \times L_1$

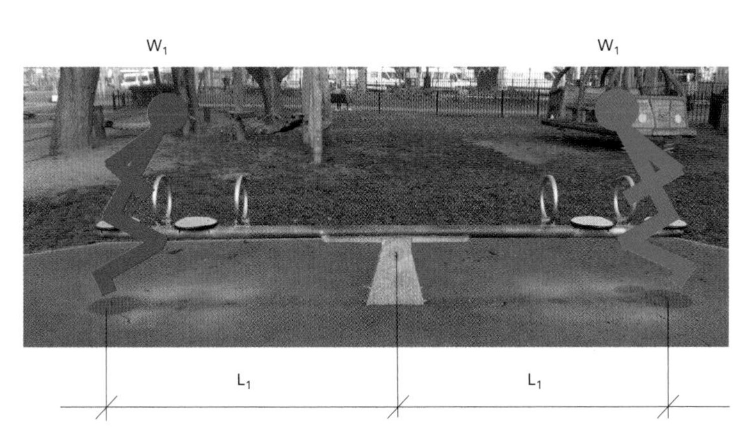

Anti-clockwise moment around pivot point,

$$M_{anticlock} = W_1 \times L_1$$

Clockwise moment around pivot point,

$$M_{clock} = W_2 \times L_1$$

If $W_1 = W_2$

it implies, $M_{anticlock} = M_{clock}$

Therefore, $\Sigma M = 0$

System is in static equilibrium

Seesaw example 1

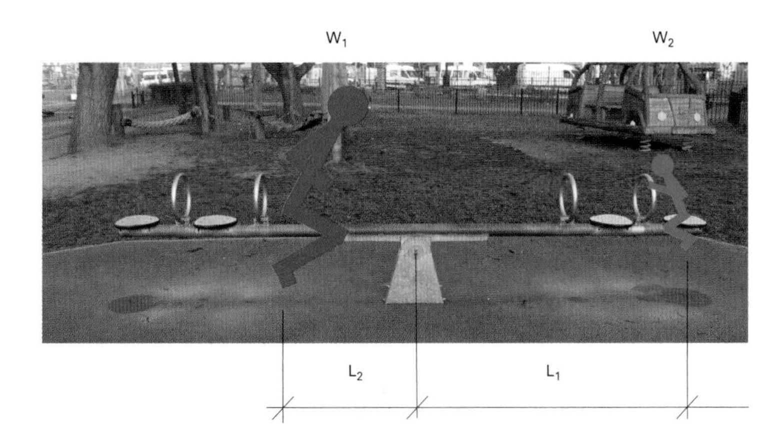

Anti-clockwise moment around pivot point,

$$M_{anticlock} = W_1 \times L_2$$

Clockwise moment around pivot point,

$$M_{clock} = W_3 \times L_1$$

If $W_1 = 2 \times W_3$

and $L_1 = 2 \times L_2$

Substituting W_3 and L_1 in clockwise moment gives:

$$M_{clock} = (W_1/2) \times 2L_2$$

$$W_1 \times L_2 = M_{anticlock}$$

it implies, $M_{anticlock} = M_{clock}$

Therefore, $\Sigma M = 0$

System is in static equilibrium

Seesaw example 2

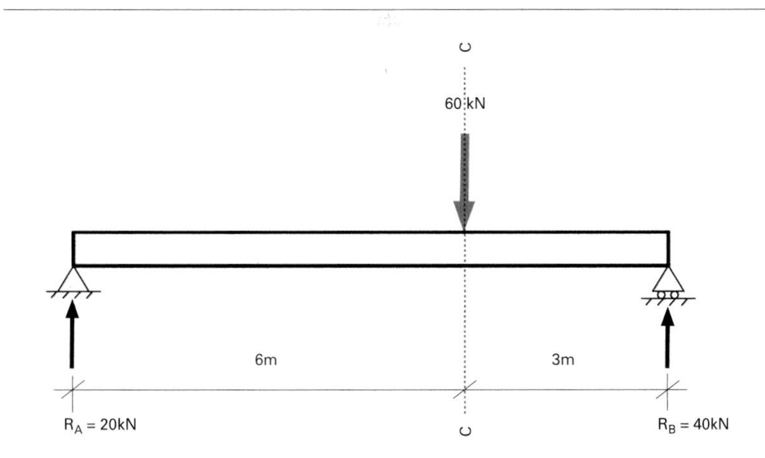

Σ vertical forces = 0

so $R_A + R_B - 60kN = 0$

Rearranging; $R_B = 60 - R_A$

& **Σ moments around a point = 0**

As Σ moments around any point are zero, it can be shown that taking the moments around the position of the point load, point C;

$$R_A \times 6 = R_B \times 3$$

Substituting R_B for R_A gives;

$$R_A \times 6 = (60 - R_A) \times 3$$

$$6 R_A = 180 - 3 R_A$$

$$9 R_A = 180$$

$$R_A = 20 kN$$

Therefore if; $R_B = 60 - R_A$

then $R_B = 40kN$

Beam example

2.1.3.6
Simple analysis

The axial force, shear force, bending moment and torsion developed in a member under various loading scenarios can be calculated with simple formulae. These member actions are often displayed graphically using force diagrams. Common member loading scenarios with the associated formulae and force diagrams are indicated on pages 36–39. In addition, an example of the 'Method of Sections' technique for determining the forces within the members of a truss is included on pages 34–35 as this explains some useful concepts of analysis and static equilibrium.

To analyse a beam accurately the support conditions must be modelled appropriately. The formulae on the following pages use the concepts of 'pinned' and 'fixed' support conditions. 'Pinned' supports act like hinges and provide no resistance to rotation, whereas 'fixed' supports are rigid and provide full resistance to rotation. A beam with pinned supports at both ends is termed 'simply supported'. A beam with fixed supports at both ends is termed 'fully fixed'.

Considering the pinned support conditions in the context of the loaded frame indicated in the diagram shown opposite bottom, it can be seen that the loaded beam cannot transfer any moment into the supporting columns. When load is applied to the beam the bottom face at midspan will experience tension while the top face will be in compression. This is termed a 'sagging' moment. The shear force applied to the beam is resisted by internal shear forces within it, which are transferred through the pinned connection into the column as axial forces.

Considering the fixed support conditions in the context of the loaded frame indicated in the diagram opposite bottom, it can be seen that no rotation between the column and beam can occur because as the beam deflects under load the column will also be forced to deflect. This alters the deflected shape of the fixed frame in comparison to the pinned frame. As with the pinned frame under load, a sagging moment is developed at the midspan of the fixed frame. Unlike the pinned frame, with the fixed frame moments also develop at the supports whereby the forces are reversed, with tension developing in the upper section of the beam and compression in the lower section. This is termed a 'hogging' moment.

The point along a fixed beam at which sagging moment turns to hogging moment (i.e. the point at which the moment is zero) is known as a point of contraflexure. Internal shearing forces are transferred through the fixed connection and into the columns as axial loads in a similar manner to pinned connections.

Fixed connections reduce the midspan bending moment and deflection of a beam significantly in comparison to pinned connections, enabling the use of smaller beams. This is demonstrated in the photographs on page 32 of a simple model of identical beams with identical loads at midspan, one with pinned and one with fixed supports. The bottom photograph clearly shows the points of contraflexure that develop on the fixed model beam and the reduced midspan deflection. The formulae on the following pages indicate that the moment at the midspan of a fixed beam under a central point load is half that of the same beam with pinned connections, and that the deflection will be four times smaller.

Another significant advantage of frames with fixed connections is their ability to resist lateral loads without collapsing, as pinned frames would. This is examined in section 2.1.7.2 on rigid framed structures.

Pinned connections are simpler to construct and less expensive than fixed connections because they are not required to resist any transferred moment and allow smaller, more slender columns to be utilized.

The concept of fixed and pinned supports is theoretical – in practice, very few connections behave as either purely pinned or rigidly fixed. These concepts are useful at the preliminary design stages to quickly assess beam and column sizes and a building's resistance to lateral forces.

Beyond the preliminary design stages connections are either designed as pinned, and the connection details are developed to accommodate a degree of rotation, or the moment transfer between the beam and column is calculated subject to the relative stiffness of the members, and the connection is designed to be capable of transferring this moment. The latter is known as a moment connection.

Beam with simply supported end conditions

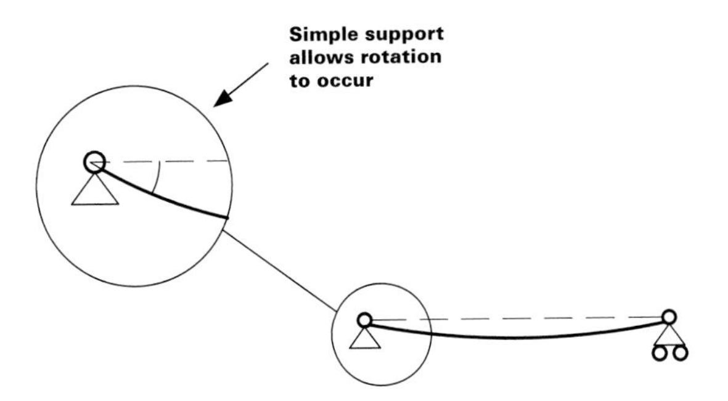

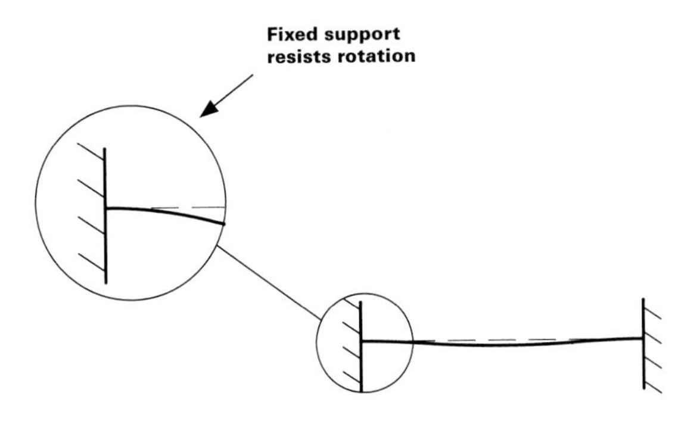

Beam with fully fixed end conditions

Simple frames with pinned and moment connection

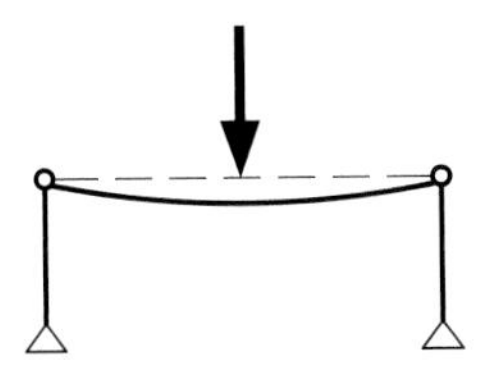

Pinned beam under vertical loading

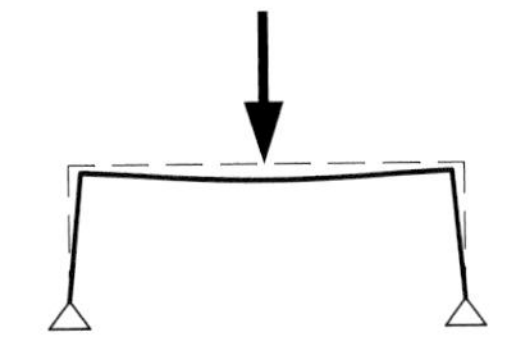

Fixed beam under vertical loading

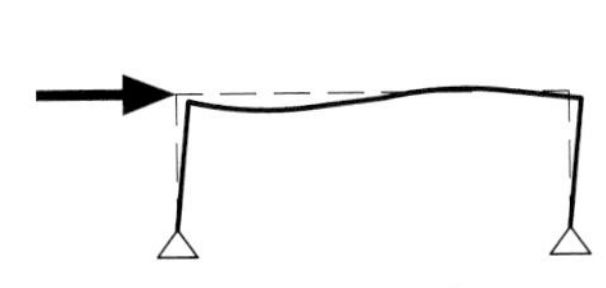

Fixed beam under lateral loading

Model of beam under load with pinned and fixed support conditions

Simple models using wood beams loaded at midspan to demonstrate the implications of 'pinned' and 'fixed' ended beam support conditions on deflection

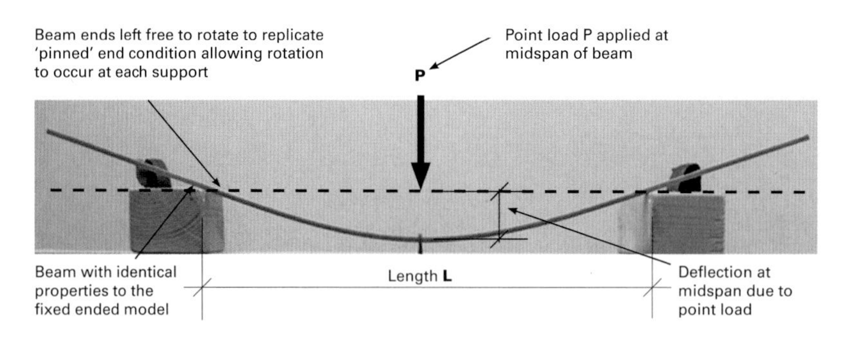

'Pinned beam'

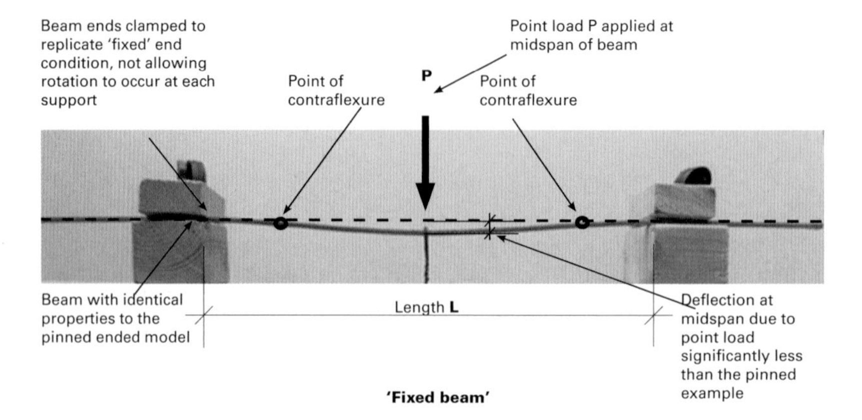

'Fixed beam'

Examples of pinned and moment connections in various materials

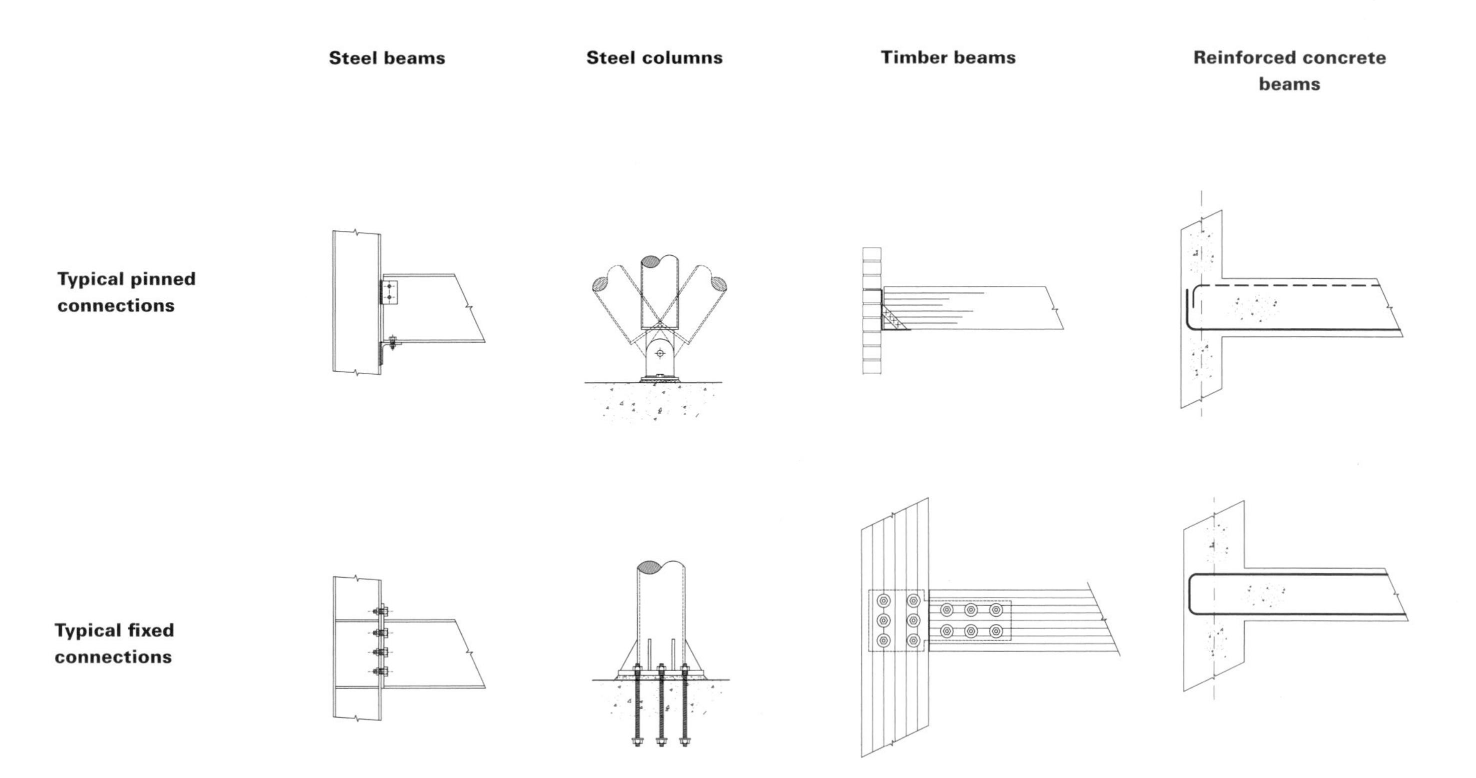

Method of sections

Glossary

F_v = vertical loads
R_v = vertical reaction
M = bending moment
F_{AB} = Axial force in truss members

The following four concepts can be used to calculate the forces in the members of a truss using the method of sections.

Concepts

i) Moment = Force x perpendicular distance from point of reference

ii) In a static system the sum of applied vertical forces equals the sum of the vertical reactions:

$$\Sigma F_v = \Sigma R$$

iii) In a static system the bending moments around any point are zero:

$$\Sigma M = 0$$

iv) Components of force:

A force can be described as two separate component forces acting at right angles to one another.

$F_x = F \cos 30\ \theta$
$F_y = F \sin 30\ \theta$

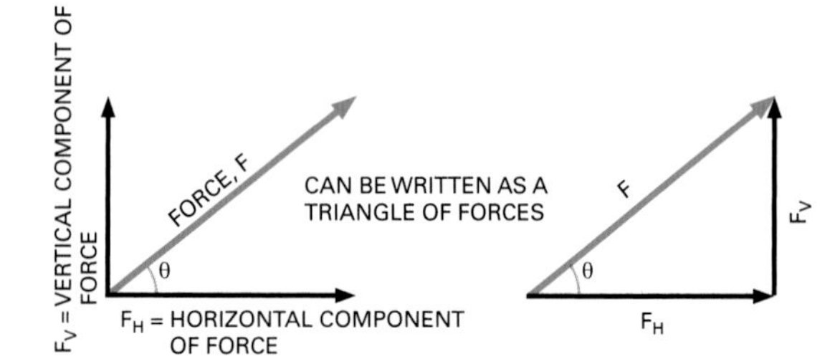

Truss with central point load

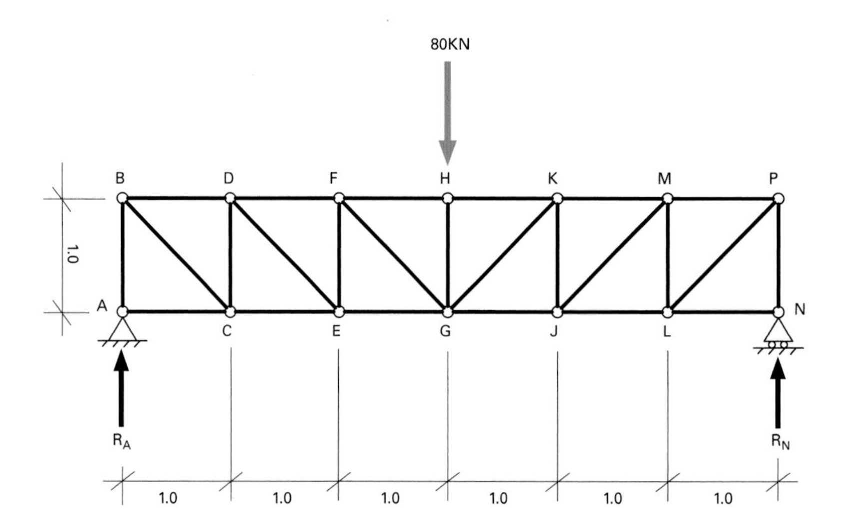

The following example shows how the Method of Sections uses the four concepts described above to calculate the forces within the vertical and diagonal members of this loaded truss.

Step 1

From Concept ii) $\Sigma W = \Sigma R$

Hence: $80kN = R_A + R_N$ (Equation 1)

From Concept iii) $\Sigma M = 0$

taking moments around support R_A: $R_N \times 6.0 = 80 \times 3.0$

substituting into equation 1 gives: $R_N = 40kN$

$R_A = 80 - R_N$

$R_A = 40kN$

Step 2

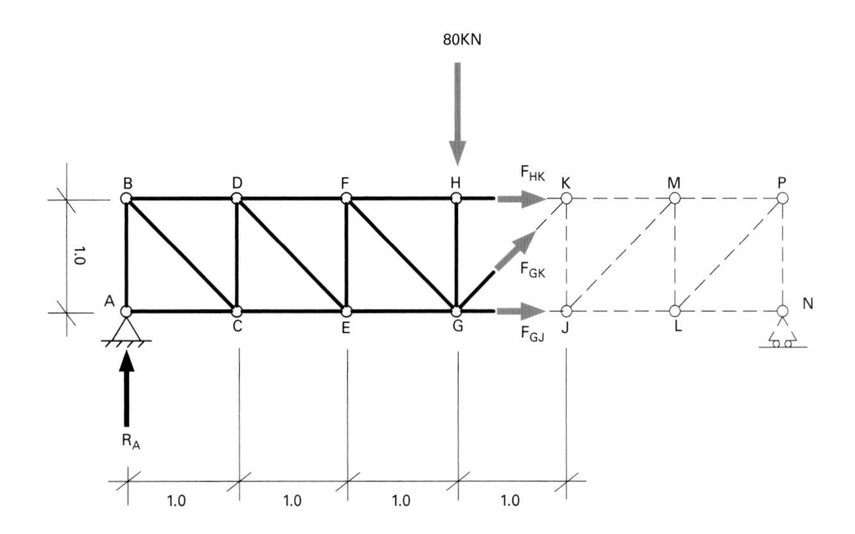

Consider the truss is cut as shown left. The forces in the individual members of the truss have to be replicated to maintain static equilibrium. Initially assume the forces act in the directions indicated (tension). Note that forces that pass through the joints produce 0 moment at these points as the perpendicular distance from line of force to point of reference is 0.

Take moment around point G (Concept iii) $(F_{HK} \times 1.0) + (R_A \times 3.0) = 0$

Substituting in R_A from Step 1 gives

$F_{HK} = -3.0\, R_A$

$F_{HK} = -40 \times 3.0$

$F_{HK} = -120$

The negative indicates that the direction of force is in the opposite direction than originally assumed, hence the force required to maintain static equilibrium in the cut truss model is compressive.

Considering moments about point K (Concept iii) $(80 \times 1.0) + (F_{GJ} \times 1.0) - (R_A \times 4.0) = 0$

substituting R_A gives: $F_{GJ} = (40 \times 4.0) - (80 \times 1.0)$

$F_{GJ} = 80$ kN

For the final unknown F_{GK} consider vertical equilibrium of the cut truss. If $\Sigma F_v = \Sigma R_v$ then the vertical component of F_{GK} plus the other vertical loads and reactions must equal zero. Where:

Vertical component of $F_{GK} = F_{GK} \sin\theta$ (Equation 2)

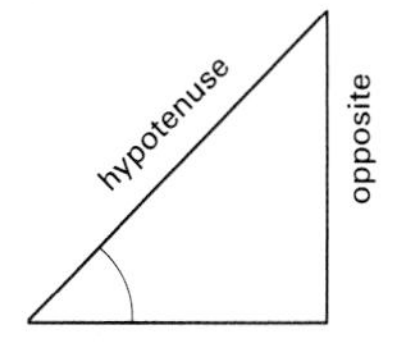

$\sin\theta = \dfrac{\text{(length of the opposite leg of the triangle)}}{\text{(length of the hypotenuse of the triangle)}} = \dfrac{1}{\sqrt{2}}$

Therefore rewriting Equation 2 gives

Vertical component of $F_{GK} = F_{GK}\,(1.0/\sqrt{2})$

Hence considering vertical forces and reactions:

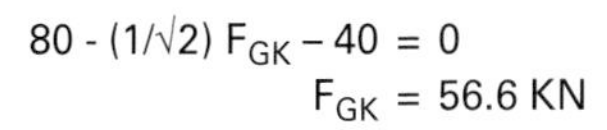

$80 - (1/\sqrt{2})\, F_{GK} - 40 = 0$

$F_{GK} = 56.6$ KN

The positive value indicates that the force F_{GK} is acting in the direction assumed on the cut truss diagram and is tensile.

This process can be repeated at adjacent nodes to calculate all the internal member forces of the truss.

2.1.3.7
Common beam formulae

Simply supported beam formulae for common load cases

W = point load (KN)
ω = uniformly distributed load (KN/m)
R = Reaction forces
L = Length (m)
I = Second moment of area (see section 2.1.5.1)
E = Young's modulus (see section 2.1.4.2)

Simply supported beam with central point load | **Simply supported beam with uniformly distributed load**

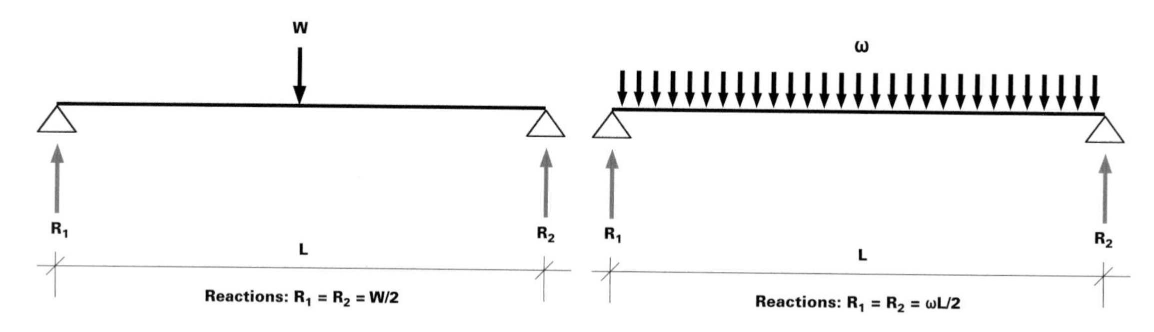

Bending moment diagrams

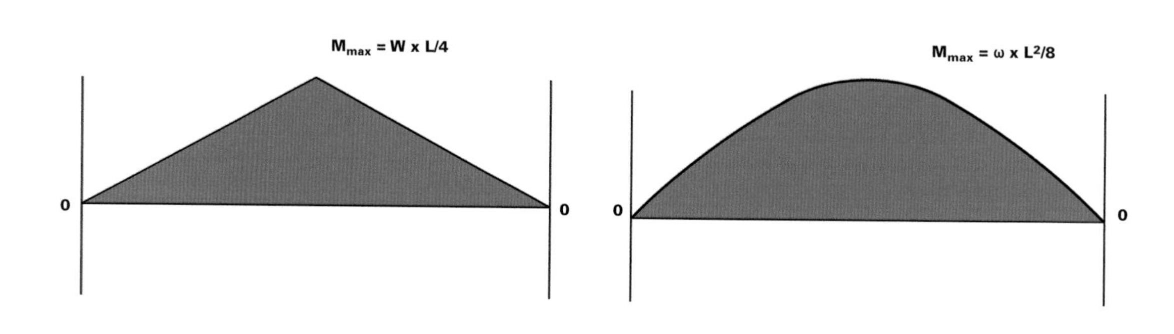

Shear force diagrams

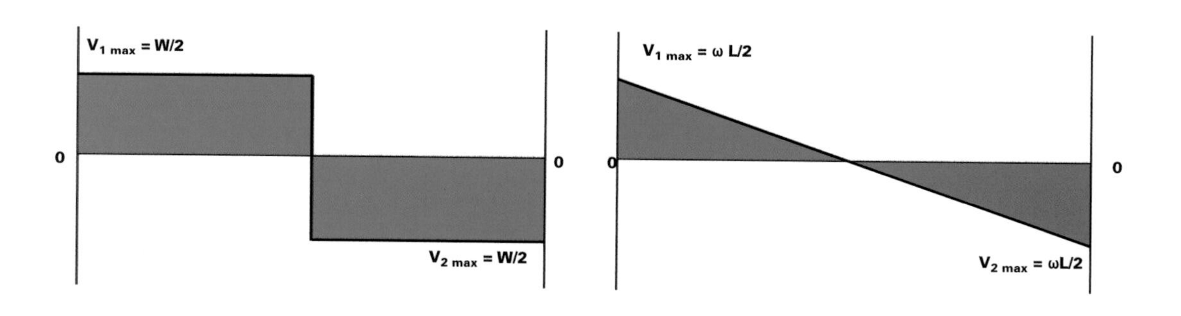

Deflection calculations

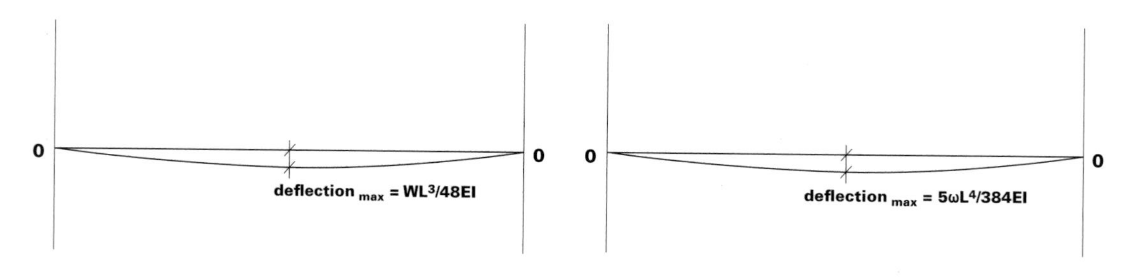

Fully fixed beam formulae for common load cases

W = point load (KN)
ω = uniformly distributed load (KN/m)
R = Reaction forces
L = Length (m)
I = Second moment of area (see section 2.1.5.1)
E = Young's modulus (see section 2.1.4.2)

Fully fixed beam with central point load | **Fully fixed beam with uniformly distributed load**

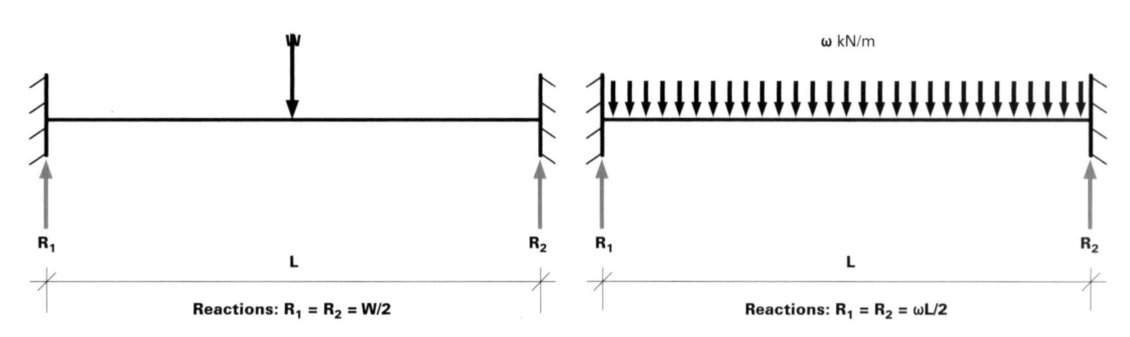

Bending moment diagrams

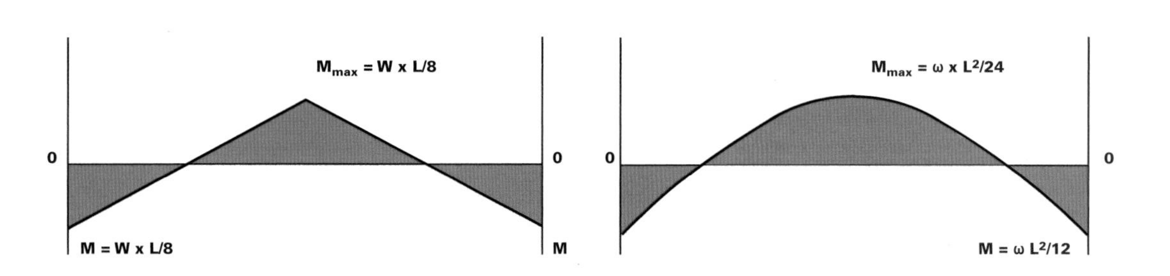

Shear force diagrams

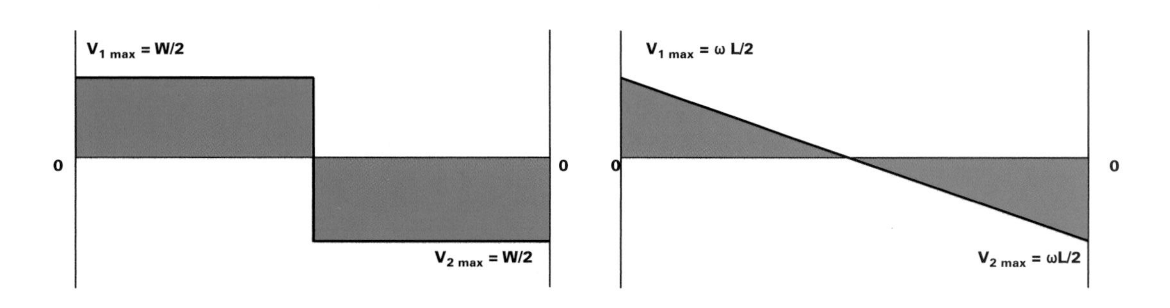

Deflection calculations

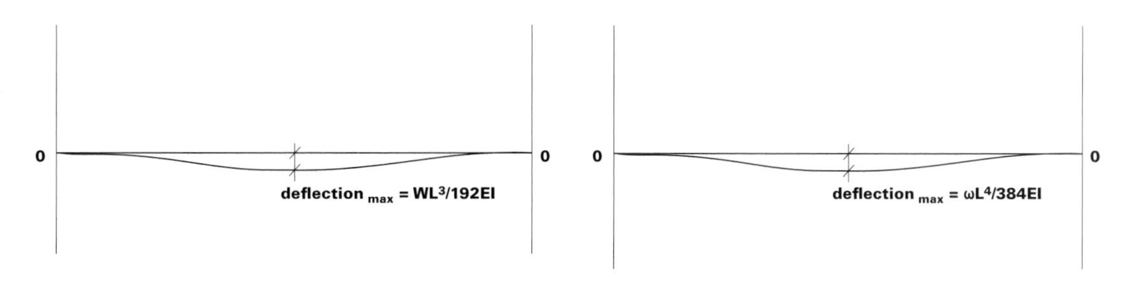

Cantilevering beam with eccentric load

W = point load (kN)
T = torsion
V = shear force
M = bending moment

R = reaction forces
$\mathbf{L_1}$ = span of fixed ended beam (m)

$\mathbf{L_2}$ = Length of cantilever

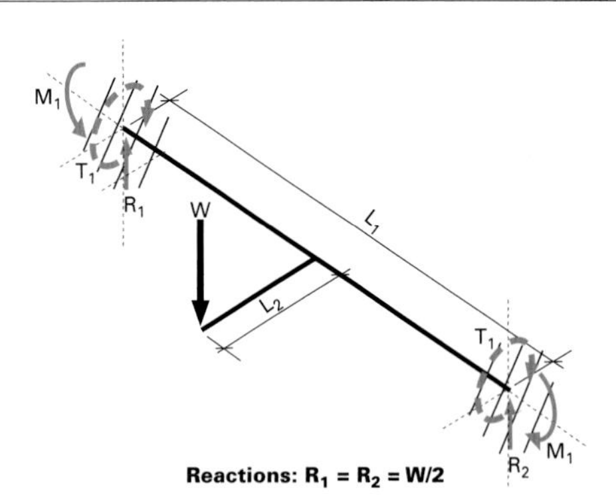

Reactions: $R_1 = R_2 = W/2$

Bending moment diagram

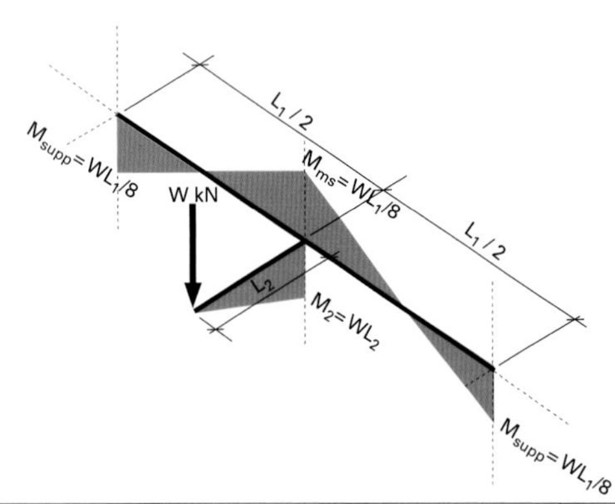

Shear force diagram

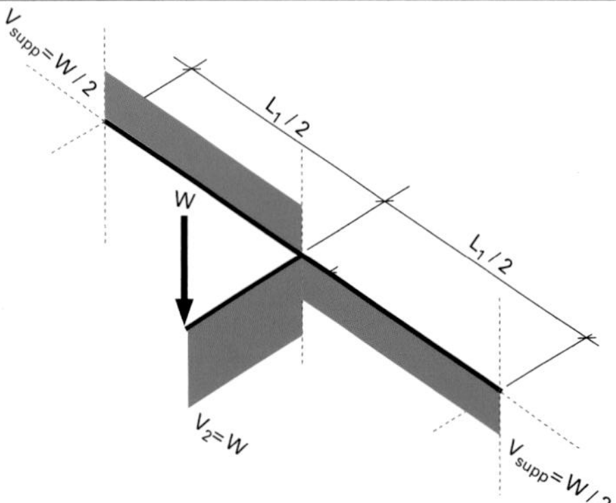

Torsion diagram

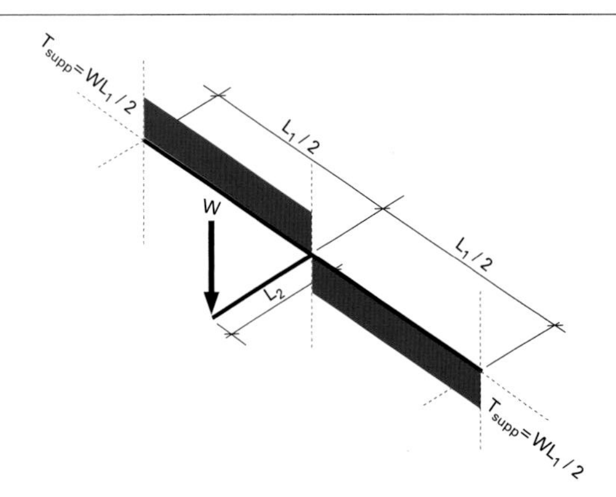

Uniformly loaded horizontal cable formulae

w = uniformly distributed load (KN/m)
L = span
h = cable sag
T = tension in cable
H = horizontal component of cable tension
V = vertical component of cable tension
s = sag ratio

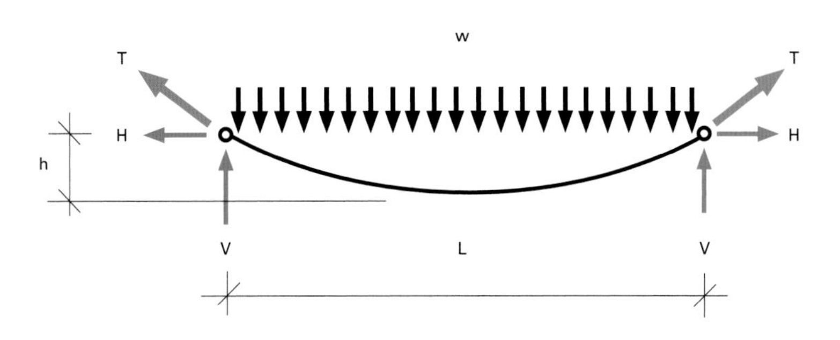

Horizontal force $H = wL^2 / 8h$
Vertical force $V = wL / 2$
Sag ratio $s = h/L$
Tension in cable $T = \sqrt{\left(\left(\frac{wL^2}{8h} + \frac{wL}{2}\right)^2\right)}$

2.1.4 Material properties

A structural component's ability to resist applied loads is based on two very basic criteria: what it is made of (material properties), and how big it is (sectional properties).

This section examines material properties; section 2.1.5 examines sectional properties.

The two most fundamental material properties that determine a material's structural characteristics are its stress and strain capacities. Stress is a measure of the force per unit cross-section of material. Strain is a ratio of the 'change in dimension' to 'original dimension' of a material when it is loaded.

2.1.4.1 Stress

External loads applied to a structural element induce internal forces within it. Stress is a measure of the intensity of these internal forces, and is expressed as force per unit area. This is normally written as N/mm2 or in Pascals (Pa) where:

$$1\ \text{N/mm}^2 = 1 \times 10^6\ \text{Pa} = 1\text{MPa}$$

As the load applied to an element increases, the internal forces and therefore the internal stresses experienced by that element increase until eventually the material reaches a limit beyond which it will fail. The limiting stress can be determined in two different ways:

i) 'Yield stress' (or 'proof stress') – this is the stress limit beyond which the material no longer behaves 'elastically' (see section 2.1.4.2).

ii) 'Ultimate stress' – this is the stress beyond which the material will fail by being either crushed or pulled apart.

The process of designing a material that will not exceed its yield stress capacity is termed elastic design because the material will behave in accordance with elastic principles in all load conditions. Materials classed as ductile, such as mild steel, can be designed to exceed their maximum yield stress using plastic design theory, which allows greater loads to be supported than elastic design. These concepts are developed further in the bending stress section and in section 2.1.4.2.

There are two types of stress that can be induced in a structural element: 'direct stress' and 'shear stress'. Direct stresses are developed when an element is subjected to an applied force parallel to its longitudinal axis. Shear stresses are developed when an element is subjected to an applied force perpendicular to its longitudinal axis.

Axial loads act parallel to the length of a member and hence induce direct stresses, whereby the magnitude of stress is calculated as the force applied divided by the cross-sectional area perpendicular to the direction of load.

Shear loads develop shear stresses on the cross-section of the loaded member in the direction parallel with the direction of load. The distribution of shear stress is termed the shear flow. This varies, the maximum occurring at the midpoint and reducing to zero at the extreme fibres. The maximum shear stress in a rectangular section is:

$$\tau_{max} = 1.5W/A$$

Where W = applied shear load
and A = cross-sectional area of member

Typically, the average stress over the member cross-section is taken as simply:

$$\tau_{average} = W/A$$

Shear forces also simultaneously induce stresses parallel with the longitudinal axis of the member called 'complementary' shear stresses. These can be explained by considering a small length of a beam under shear load as illustrated on the opposite page. In order for the small length of beam to maintain static equilibrium there must be an additional pair of equal and opposite forces acting at right angles to the main shear forces in the beam. In some materials, including timber, these complementary shear stresses can be more critical than the main shear stresses.

Bending forces, as with axial forces, induce direct stresses within an element. Unlike axial force-induced stresses, the magnitude and the direction of direct stresses due to bending vary across the cross-section of a member. The extreme fibres of an element under bending experience the highest tension and compression stresses simultaneously. In between the extreme fibres the stress levels reduce to a point where stress is zero. This point is known as the neutral axis. In accordance with elastic theory, bending stress in a beam is calculated by dividing the applied bending moment by the 'section modulus' of the beam. This is a sectional property explained in section 2.1.4.2.

Elements under axial stress

External compressive point load applied to column

External tensile point load applied to column

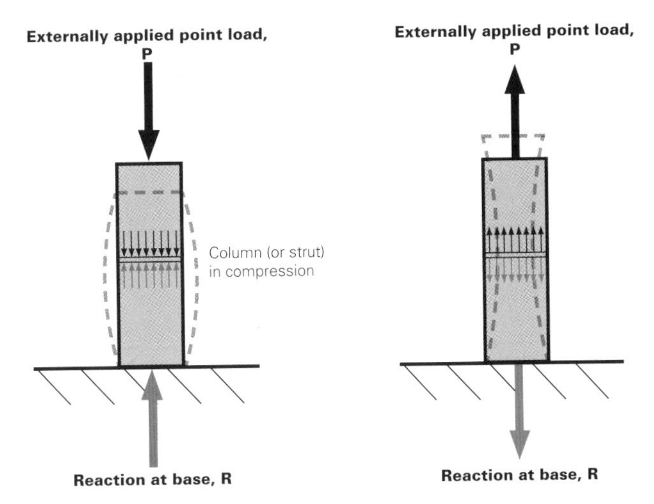

Stress, $\sigma = P/A$

where P = applied axial force (in Newtons, where 1kg = 10N)
and
A = cross-sectional area of the element

Shear stress in beam under bending

External point load applied to beam

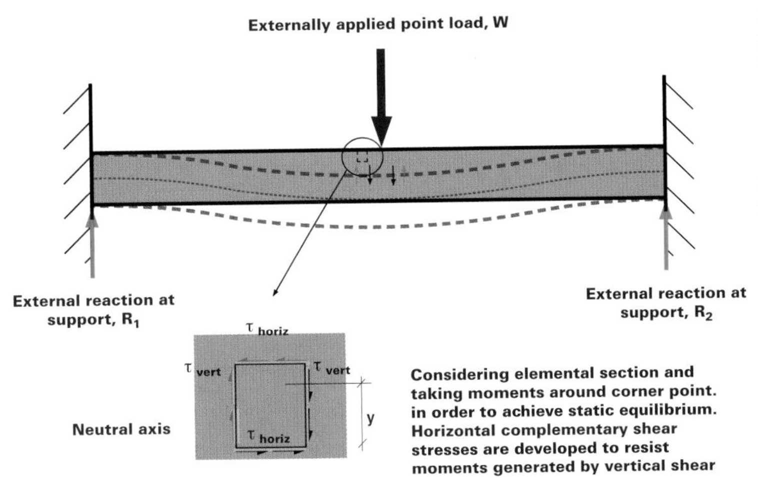

Considering elemental section and taking moments around corner point. in order to achieve static equilibrium. Horizontal complementary shear stresses are developed to resist moments generated by vertical shear stresses.

Average vertical shear stress on cross-section perpendicular to longitudinal axis:

Shear stress, $\tau_{vert} = W/A$

W = applied shear force (N)
A = cross-sectional area (mm^2)

Complementary shear stress on cross-section parallel to longitudinal axis:

Shear stress, $\tau_{horiz} = WA'y/bI$

W = Applied shear force
A' = Sectional area of section considered
y = Distance from centroid of area A' to elastic neutral axis
b = Breadth of element considered
I = Second moment of area of whole section (see section 2.1.5.1)

Bending stress in beam under bending

Beam under bending showing compression/ tension and neutral axis

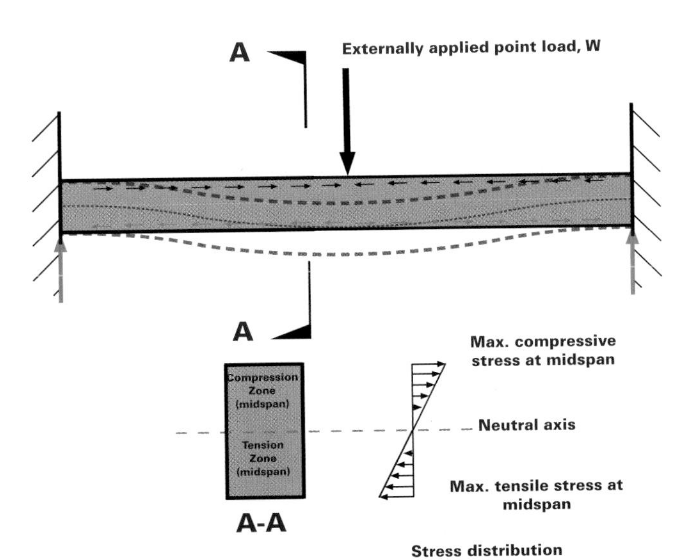

Bending stress, $\sigma = M/Z$

Where: M = bending moment
Z = section modulus
(see section 2.1.5.1)

UNIVERSITY
OF SHEFFIELD
LIBRARY

Torsional forces induce shear stresses in the plane perpendicular to the longitudinal axis of a member. These shear stresses act in a circular nature and their magnitude varies linearly across the cross-section from zero at the centre to a maximum at the outer face. Due to the radial nature of torsional stresses, the 'shear flow' is highly dependent on the shape of the section under stress. Solid circular sections and hollow sections have a closed circular route that stress can follow and hence these shapes are able to resist torsional loads more efficiently than 'open' sections (such as steel I beams). The torsional stress is calculated using the polar second moment of area, which for a solid circular section is:

Polar second moment of area $J_{solid\ circle} = \pi D^4/32$

Torsional stress $\tau_{solid\ circle} = Tr/J$

Where T = torsion
D = diameter of shaft
and r = distance from centre to point considered

For τ_{max}, r = radius of section

The polar second moment of area of a rectangular section is more complicated and beyond the scope of this book; however, the stress in a rectangular section is often approximate to:

$$\tau_{solid\ rectangle} = 2T/h_{min}^2(h_{max} - h_{min}/3)$$

where h_{max} and h_{min} represent the breadth and width of the rectangular cross-section.
The values of the direct yield stress capacity, σ, and the shear yield stress capacity, τ, are generally not the same for a given material. For example, for mild steel:

Normal tensile yield stress σ = 275 N/mm²

Normal compressive yield stress σ = 275 N/mm²

Shear yield stress τ = 165 N/mm²

Different materials can withstand differing maximum shear and direct stress values, making some more suited to structural applications than others. A list of various common materials with their associated stress capacities is provided in the table on page 46.

Element under torsion

Eccentric external point load inducing torsional stress in beam

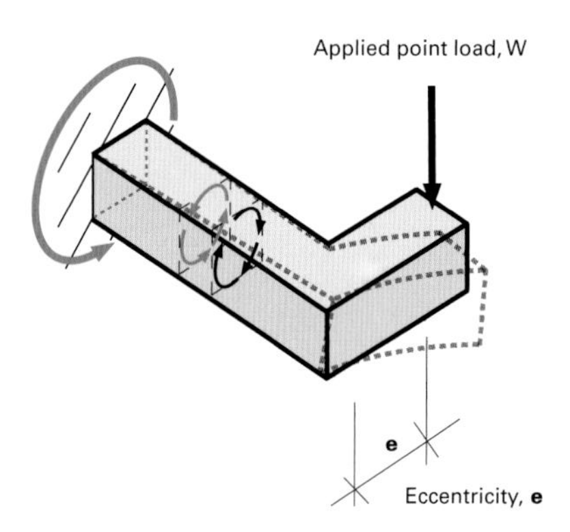

Approximate shear stress due to torsion in solid rectangular section:

$$\tau = 2 \times T / h_{min}^2 (h_{max} - h_{min}/3)$$

Where h = dimensions of rectangular cross-section (m)
T = applied torsion = We (kNm)

Shear stress due to torsion in solid circular shaft:

$$\tau = Tr/(\pi D^4/32)$$

Where T = applied torsion = We (kNm)
D = diameter of shaft (m)
r = distance from centre of shaft to point of measurement (m)

Metals and concrete are 'isotropic' materials, meaning that they have identical material properties in all directions. Hence a cube of concrete or metal will support the same compressive load regardless of which face of the cube the load is applied to. The same is true for tensile and shear loads. Timber and carbon fibre are orthotropic materials, meaning that their material properties vary in different axes. For example, a cube of timber will compress more easily when the load is applied perpendicularly to the grain than if it is applied parallel to it. In addition, the shear stress capacity of timber parallel to the direction of its grain is significantly less than the shear strength perpendicular to it. Because of this the complementary shear stresses described previously are often the critical shear design criteria of a timber beam under vertical load as opposed to the main shear stresses, which act in the direction of the applied load. Hence, when designing in orthotropic materials the orientation of the material laminations has to be considered at the design stage.

While both concrete and mild steel are isotropic materials they differ from one another in that the tensile and compressive strength of mild steel are identical. Concrete, however, has a high compressive capacity but negligible tensile strength in all axes, primarily owing to the microscopic cracks that develop in it during curing. Bending moments develop simultaneous compressive and tensile stresses in a structural member, and hence a concrete element would fail under very small loads due to its poor tensile capacity. To counter this, concrete is reinforced with longitudinal steel reinforcing bars in areas that are subject to tensile forces.

Reinforced concrete beam section

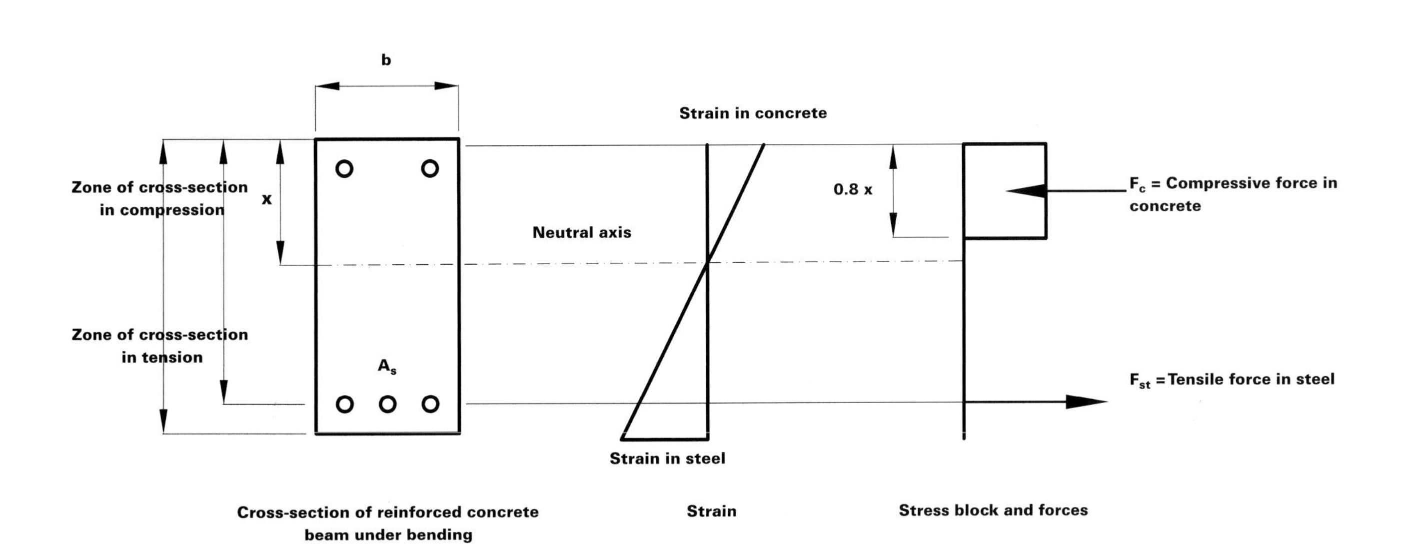

Cross-section of reinforced concrete beam under bending

Strain

Stress block and forces

2.1.4.2
Strain

When a sample of material is placed under load it will undergo some deformation. This deformation will be either via elongation, compression or shearing depending on how the load is applied. Strain is a measurement of the ratio of the extent of deformation under load against the original dimension of a sample of material. There are several different types of strain including linear, volumetric and shear. Linear strain is the ratio of the elongation under axial load against the original length. This is written as:

Strain, $\varepsilon = dl / l$

where dl = extension, and
l = original length

When they are loaded, most materials exhibit 'elastic' behaviour in accordance with Hooke's Law. As load is applied materials deform and when the load is removed they return to their original dimensions. Plotting the strain against the stress in a material as it is loaded produces the graph illustrated on the opposite page. This example is based on a mild steel sample. The straight line area indicates the linearly elastic region. In this area the material adheres to Hooke's Law and returns to its original size as load is released. The ratio of stress divided by strain in this region is a constant value known as the elastic modulus. For tensile forces that induce tensile stresses and strains, this is more commonly known as Young's modulus. Other elastic moduli include the shear modulus, volumetric modulus and Poisson's ratio, all of which are briefly explained in the diagrams opposite.

Young's modulus, E = linear stress/linear strain
$= \sigma / \varepsilon$

This value, combined with other sectional properties, is used to calculate the 'stiffness' of structural members using the formula:

Stiffness $K = EI/L$

Where I = second moment of area (see section 2.1.5.1)
L = length of member

The stiffness of a member or system of members is used when calculating the deflections of structural members and to determine the amount of load that is resisted by each of the members in a system where the stiffer elements will attract the greater loads.

As the load applied to a sample of material is increased it will eventually reach its elastic limit beyond which it will not return to the exact dimensions upon release of the load. At this point the material is behaving 'plastically' and is represented by the plastic range indicated on the graph opposite.

As load, and therefore stress, is increased incrementally the material will eventually reach its ultimate stress capacity, at which point it will break.

Both mild steel grade S275 and Aluminium 6061-T6 have very similar stress capacities of around 275 Newtons per square millimetre, meaning that they will be able to support very similar loads prior to reaching their yield strengths. Aluminium 6061-T6, however, has a Young's modulus of 68,900 Newtons per square millimetre, which is approximately three times smaller than that of mild steel at 205,000 Newtons per square millimetre; hence an aluminium beam will deflect three times more than an identical-sized mild-steel beam under the same loads. In this example the steel beam can be said to have a 'flexural stiffness' three times greater than an aluminium beam as the geometrical properties I and L are constant.

The extent of deformation that a material is able to undergo before failure occurs determines whether it is classified as 'brittle' or 'ductile'. Materials that fail before strain reaches 5 per cent are classified as brittle. These include concrete, timber, glass and ceramics. Brittle materials tend to fail suddenly and without warning. Materials such as mild steel and aluminium are classified as ductile, as they can exhibit a significant degree of deformation prior to failure. This can often be seen as a change to the cross-section of an element in tension, or high deflection of beams in bending.

Other properties that affect the performance of the most common structural materials are included in the following section.

Types of strain

L = original dimension
dL = extension
W = force

Tensile strain **Shear strain** **Volumetric strain**

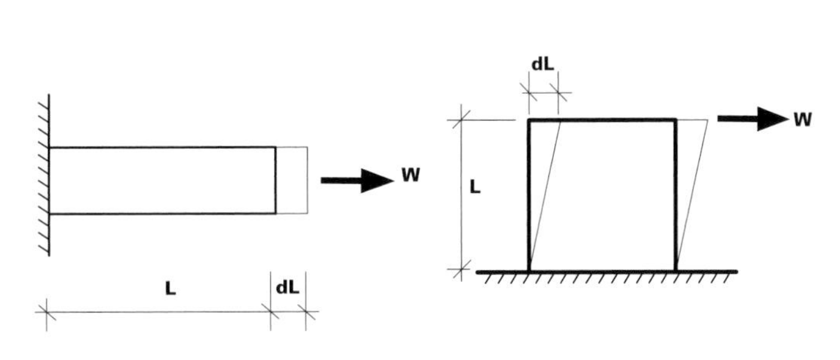

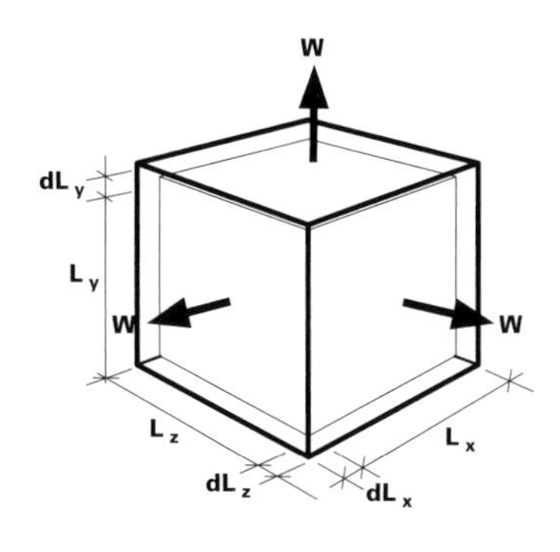

Tensile strain, $\varepsilon_L = dL/L$

Tensile stress, $\sigma = W / A$

Tensile modulus, or Young's modulus:

$E = \sigma / \varepsilon$

Shear strain, $\varepsilon_s = dL/L$

Shear stress, $\tau = W / A$

Shear modulus, or modulus of rigidity:

$G = \tau / \varepsilon$

Volumetric strain, $\varepsilon_v = dL_x/L_x + dL_y/L_y + dL_z/L_z$

Volumetric modulus, or bulk modulus:

$K = dp / (dV/V)$

where: dp = differential change in pressure on object
dV = differential change in volume of an object
V = initial volume of object

Poisson's ratio

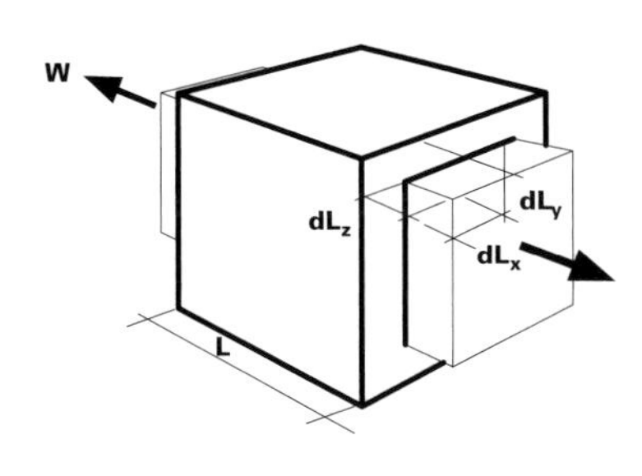

Poisson's ratio, ν = transverse strain / longitudinal strain

$\nu = (3K - 2G)/(6K + 2G)$

$E = 2G(1 + \nu)$

$E = 3K(1 - 2\nu)$

Stress/strain graph

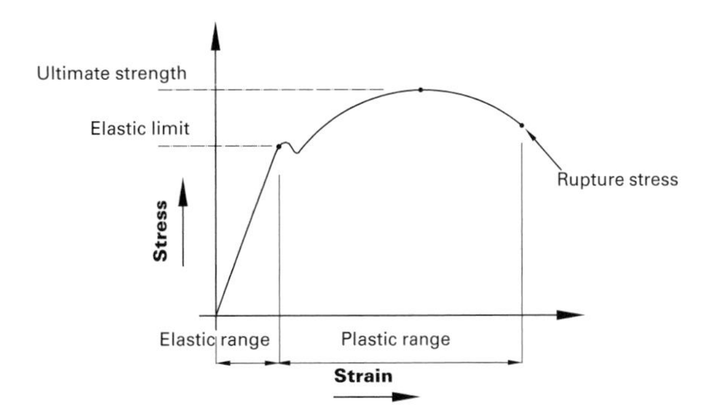

Young's modulus

The yield and ultimate stress limits for some typical materials, together with their associated Young's moduli

Material	Yield stress (N/mm²)	Ultimate stress (N/mm²)	Young's modulus (N/mm²)
Mild steel (ASTM A-36)	275	420	205,000
High-strength steel (ASTM A-514)	690	760	205,000
Aluminium 6061-T6	276	380	68,900
Iron	100	350	211,000
Titanium	225	370	120,000
Tungsten	550	620	410,000
Concrete (C40)	N/A	40 (compressive)	30,000
Human hair	N/A	380	3,890
Glass	N/A	33	70
Carbon fibre (Cytec Thornel T-650/42 12KL)	N/A	4,820 (tensile)	290,000
Bone (femur)	N/A	135 (tensile) 205 (compression)	17,000
Natural rubber	N/A	28 (tensile)	10
Graphene	N/A	130,000	1,000,000
Douglas fir (softwood)	N/A	50	12

2.1.4.3
Steel properties

Grade

Structural steel is graded to identify its yield stress characteristic. The most common grades for steel sections in the UK are S275 or S355, which represent a yield stress capacity of 275N/mm2 and 355N/mm2 respectively. Higher-strength steels contain higher levels of carbon. While increasing the carbon levels adds strength, it also increases brittleness and makes steel less easy to weld. More brittle steel has a greater susceptibility to brittle fracture in cold conditions, and hence steel must be specified not only based on its yield stress characteristics but also the climate conditions to which it will be exposed.

Brittleness can be assessed by measuring the resistance of steel to impact. A common test to assess impact resistance is the Charpy v-notch test, which involves using a pendulum to strike a sample of material and calculating the energy absorbed in the sample by measuring how far the pendulum swings back after striking the sample.

In accordance with the code, EN10025: Part 2:2004, steel is specified as follows:

S	**275 JR**
S	denotes that it is structural-grade steel
275	denotes the yield stress and can vary but is most often 275 or 355
JR	denotes the resistance to impact and can vary from JR, J0, J2 to K2, where JR is for normal conditions and K2 for severe exposure conditions

Fatigue

Under cyclic loading and unloading, metal structures can develop microscopic cracks at the surface owing to 'fatigue'. If left to develop, fatigue cracking can lead to the sudden catastrophic failure of a member. Structures subject to cyclic loading – such as road bridges, certain industrial buildings, gymnasiums and dance floors – must be designed against fatigue failure. This is done by estimating the number of loading cycles over the lifetime of the structure and using experimental data to reduce the design stress of the steel.

2.1.4.4
Concrete properties

Grade

Concrete is graded in terms of its compressive strength and the exposure conditions that it will be subject to. The actual design-mix proportions, including the percentage of cement, will then be designed specifically to meet these requirements. In reinforced concrete the cover of the concrete to the steel reinforcing bars is also an important parameter. The 'cover' must be sufficient to ensure the steel reinforcement is not exposed to any chemicals or water in the environment that could cause it to rust. As steel rusts it expands. This causes the concrete to spall, which in turn leads to greater damage occurring. Minimum depths of cover are provided in the various concrete codes; they generally range from 20mm to 75mm depending on the severity of the exposure conditions.

Shrinkage

Concrete can shrink in several different ways after it is poured, owing to the loss of moisture and subsequent change in volume. These ways include drying shrinkage, plastic shrinkage, 'autogenous' shrinkage and 'carbonation' shrinkage. All types of shrinkage form cracks, which can affect the durability and appearance of the material. Shrinkage can be controlled in several ways. These include reducing the size of the concrete pour, protecting curing concrete from drying out by covering it with wet cloth, or reducing the volume of water in the concrete mix by using chemical additives called plasticizers.

Creep

Creep is a phenomenon whereby a solid material under constant load gradually deforms. As concrete beams are loaded they are subject to creep, which results in a gradual increase in deflection over time. The degree of creep is subject to many criteria including the concrete mix design and the relative humidity during curing and in-use conditions. In certain circumstances long-term creep deflections can be up to twice the short-term dead load deflections. The implications of creep can be particularly significant for concrete beams spanning over glass façades or non-load-bearing partitions. In these situations as the deflection of the concrete beam increases the non-load-bearing elements can be subjected to load that they are not designed to support, causing damage to occur. The effects of creep are allowed for in the design process by reducing the Young's modulus of the concrete by up to two-thirds at the design stage.

2.1.4.5
Timber properties

Strength class
Timber can be chosen by species, but is more commonly specified by stress grade for building purposes. Each piece of timber with a particular strength class will have similar bending, compression and shear capacities.

Orthotropic
Timber is an orthotropic material, and has varying structural properties in different directions. This is particularly relevant in shear design of timber beams as the shear capacity parallel to the grain is significantly lower than the shear capacity perpendicular to the grain; hence, when a beam is loaded in direction perpendicular to the grain it will generally fail in shear due to the complementary shear stress (see section 2.1.4.1) as opposed to the normal shear stress.

Natural material
Timber being a naturally occurring material means that it contains imperfections and irregularities such as knots and can develop splits, known as shakes, as it dries out. In addition timber is a hygroscopic material meaning that it will give up moisture as it dries out or take up moisture from the atmosphere depending on the relative humidity of its surroundings. Timber is unique among structural materials in these respects. Engineered timber products, such as glulam, Laminated Veneer Lumber (LVL) and Cross Laminated Timber (CLT) are manufactured from thin layers of timber glued together. This ensures enhanced mechanical properties in comparison to standard timber, as any imperfections are distributed across the length of the member as opposed to being concentrated in one area at, say, a knot. Engineered products also are more dimensionally stable as the thin veneers can be dried effectively during the fabrication process, thus alleviating the issue of drying out while in use.

Creep
As with concrete, timber is subject to creep; the strain can increase by 60 per cent over ten years under permanent load. This is often allowed for in the design by the use of load duration factors, with higher factors being applied to loads applied for longer periods.

Service class
Timber exhibits different properties when wet, and therefore the design must recognize the likelihood of the timber becoming wet and amend the material properties accordingly.

Properties	**Strength classes**					
Stress	C14 MPa	C16 MPa	C18 MPa	C22 MPa	C24 MPa	C27 MPa
Bending	4.1	5.3	5.8	6.8	7.5	10.0
Tension	2.5	3.2	3.5	4.1	4.5	6.0
Compression parallel to grain	5.2	6.8	7.1	7.5	7.9	8.2
Compression perpendicular to grain	1.6	1.7	1.7	1.7	1.9	2.0
Sheer parallel to grain	0.6	0.6	0.6	0.7	0.7	1.1
Modulus of elasticity	MPa	MPa	MPa	MPa	MPa	MPa
*E*mean	6,800	8,800	9,100	9,700	10,800	12,300
*E*minimum	4,600	5,800	6,000	6,500	7,200	8,200

1MPa=1N/mm^2

2.1.5 Sectional properties

The dimensions of a structural element's cross-section significantly affect the ability of that member to resist applied loads.

The following sections explain the relationship between geometrical sectional properties and the axial and bending stress capacities of an element.

They are not intended to give detailed guidance on the design of structural elements, as that is beyond the scope of this book, but rather to provide a conceptual understanding of how cross-sectional geometry can impact on the behaviour of structural elements.

2.1.5.1 Bending

A simply supported beam with a vertical load placed at midspan will develop a bending moment. The upper fibres of the beam at midspan will experience a compressive stress while the lower fibres will experience a tensile stress. The stress across the cross-section of the beam between these extreme fibres will vary as described in section 2.1.4.1. At a particular position on the cross-section the stress will be zero. This is known as the neutral axis.

Intuition dictates that a 30-centimetre ruler orientated as shown in the photograph opposite left will be 'harder', i.e. require a greater load in order to bend than the ruler orientated as shown in the photograph on the right.

This apparent increased strength of the ruler in the photograph on the left is due to a geometric property called the 'second moment of area'. If a cross-section is divided into a series of smaller areas and each of these areas is multiplied by the square of the distance from their centroid to the neutral axis, the summation of these quantities for the whole cross-sectional area is the second moment of area. For a rectangular section, this is calculated as:

Second moment of area, $I = BD^3 / 12$

where B = breadth of section, and
D = depth of section

Bending theory relates second moment of area, bending moment and stress in the equation:

$$M / I = \sigma / y$$

where M = bending moment,
I = second moment of area,
σ = bending stress, and
y = distance to neutral axis

Written alternatively:

$$\sigma = M y / I$$

This is commonly rewritten as:

$$\sigma = M / Z_{el}$$

where Z_{el} is called the 'elastic section modulus'.

Hence, looking back at the intuitive example of the plastic ruler it can be seen that with a rule with cross-sectional dimensions of 35 x 3 millimetres thick, the associated elastic section moduli in the two different rectilinear orientations are as indicated opposite.

Many building design codes are written using plastic, as opposed to elastic, design theory. Elastic design limits the maximum stress within a section to the elastic yield or proof stress. A beam designed in accordance with elastic theory will reach its maximum bending capacity when the extreme fibres on its upper and lower faces reach their elastic stress limit, as indicated in the stress distribution diagrams opposite. The elastic section modulus explained above is valid when this triangular stress distribution exists.

Plastic design allows for some plastic deformation of the extreme fibres of a beam in, for example, bending to occur when they reach the elastic stress limit, thus distributing stress to the lower fibres, which can then also be designed to develop full elastic stress capacity. A stress block indicating a section that has developed full plastic capacity is indicated opposite.

The elastic section modulus, Z, can be replaced with the plastic section modulus, S, in the equations above to calculate the maximum plastic moment capacity of a section.

The plastic section modulus of a rectangular beam is:

$$S = BD^2/4$$

A rule orientated in two directions

Cross-section of rule considered in two different orientations

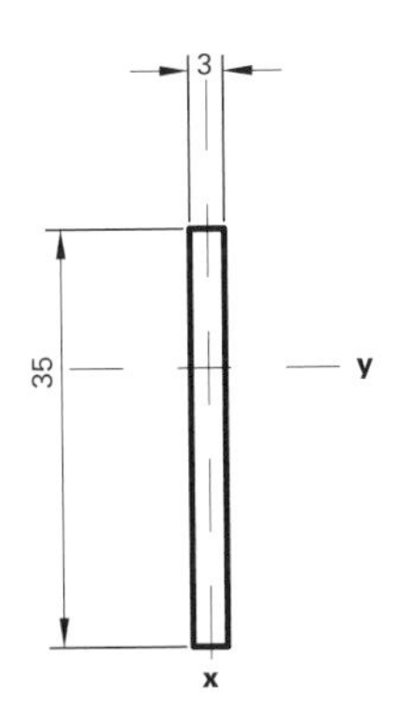

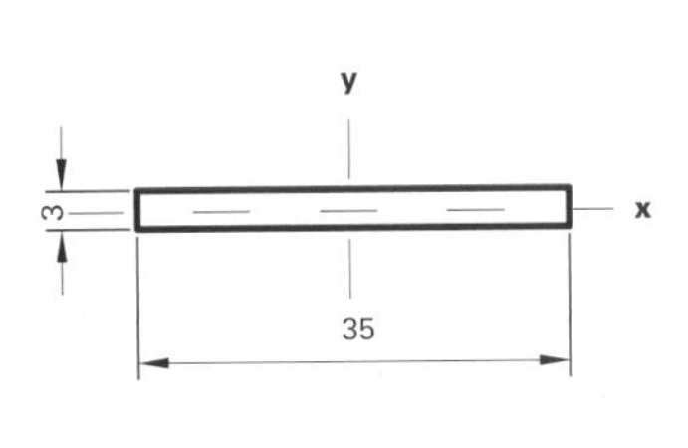

Cross-sections of rule with section moduli

Elastic section modulus about the x–x axis:

Elastic section modulus about the y–y axis

Elastic section modulus about the x–x axis:

$Z_x = BD^2/6$

Where B = 3mm
D = 35mm

Hence
$Z_x = 3 \times 35^2/6$
$= 612.5$ mm

When D = 35mm the elastic section modulus of the section is over 11 times higher than when the section is rotated through 90°. Providing the compression flange of the section is restrained this equates to the deeper section being capable of supporting over 11 times more load when orientated with a greater depth.

Elastic section modulus about the y–y axis:

$Z_y = BD^2/6$

Where B = 35mm
D = 3mm

Hence
$Z_y = 3^2 \times 35/6$
$= 52.5$ mm

Stress distributions across cross-section of beam in bending

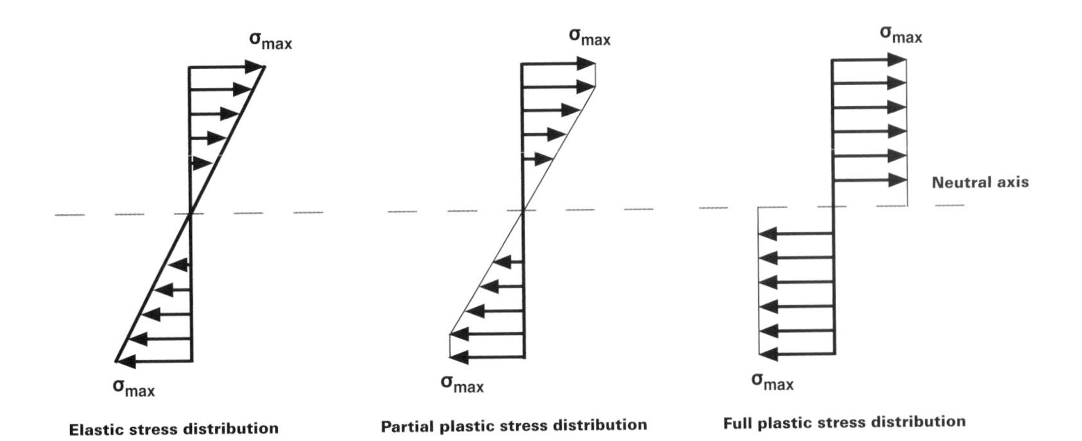

2.1.5.2
Axial compression

Members under axial compressive load can fail in two fundamentally different mechanisms:

i) Compressive failure
ii) Buckling

Compressive failure is a function of the cross-sectional area of the section and the strength of the material. Quite simply: if the load applied is too great for the column to withstand, it will crush the member.

Hence, capacity of strut owing to pure compressive load is as follows:

$$P_{comp} = A\,\sigma_{compcap}$$

where P_{comp} = crushing capacity of the strut under pure compressive load
A = area of section
$\sigma_{compcap}$ = compressive stress capacity of column material

Compressive failure generally governs the design of 'short' columns. Longer, more slender, columns, however, are prone to fail via the second mechanism, buckling, which occurs before they reach their ultimate compressive capacity.

When a long slender column is placed under an increasing axial load, that column will be seen to start to bow or buckle at a certain load magnitude. This can be demonstrated simply with a 30-centimetre plastic ruler as it is loaded carefully by hand.

If the load is increased further, the column will eventually fail in buckling rather than crushing. Buckling, unlike compressive failure, is a function of both the height and the sectional properties of a member.

As the slenderness of a column increases the criteria governing its axial strength alters from a stress-governed crushing capacity to a geometry-governed buckling capacity once the slenderness exceeds a certain limit. The graph below indicates this relationship.

While the height of a column is simply the distance from the base to the top, the 'effective height' is the extent of the column that is subjected to buckling and is determined by the restraint conditions at either end. Pinned supports at the top and bottom provide no restraint to rotation, and therefore the deflected shape of the column will be a single curve as it is loaded axially, as indicated in the photograph of a rule opposite. When the top and bottom supports are fixed, however, no rotation can occur at these points and the deflected shape of the axially loaded column will change. The length over which buckling occurs in a pin ended column is half of the length over which buckling can occur in a fully fixed column. A column with one end fixed and one end free to rotate and move (a cantilever) will have an effective buckling length of twice a pinned column. These and other end restraint conditions together with the associated column effective lengths are demonstrated graphically on page 54. Equations developed by Euler describe the critical loads columns can withstand prior to buckling. These are:

critical buckling load, P_{comp}

$$= \frac{\pi^2 EI}{Le^2}$$

critical buckling stress, θ

$$= E\left(\frac{\pi^r}{Le}\right)$$

where P_{comp} = compressive load in column
E = Young's modulus
Le = effective length
r = radius of gyration (see below)

The 'radius of gyration' is a geometrical property, where:

$$r = \sqrt{(I/A)}$$

where I = second moment of area (as explained in section 2.1.5.1)
A = Cross-sectional area

Slenderness is defined as:

$$\lambda = L_e/r$$

where L_e = the effective length of the column
r = least radius of gyration

As can be seen from the equations above the buckling capacity of a column is inversely proportional to the effective length of the column squared. Therefore, doubling the effective length will reduce the buckling capacity by a factor of $2^2 = 4$. In the case of a non-symmetrical column section, the second moment of area will be different depending on which axis is being considered. As shown in the example of the 30-centimetre ruler, slender columns always fail in buckling around their weakest axis and hence the slenderness must always be calculated on the basis of the minor, or smaller, axis. For this reason, typical column sections such as standard steel universal columns (UC) tend to be relatively symmetrical in comparison to, for example, universal steel beams, which have large disparities between their slenderness ratios in the x and y axes (see diagrams on page 54).

A 30-centimetre rule under load by hand

Axial load applied to a plastic rule

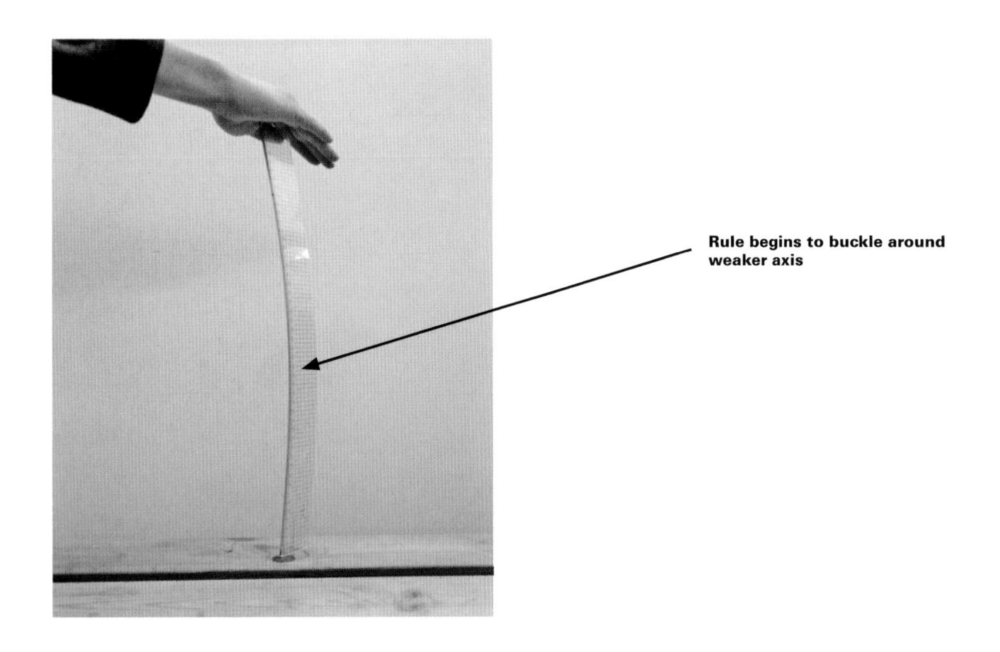

Column compression and buckling

Compressive capacity

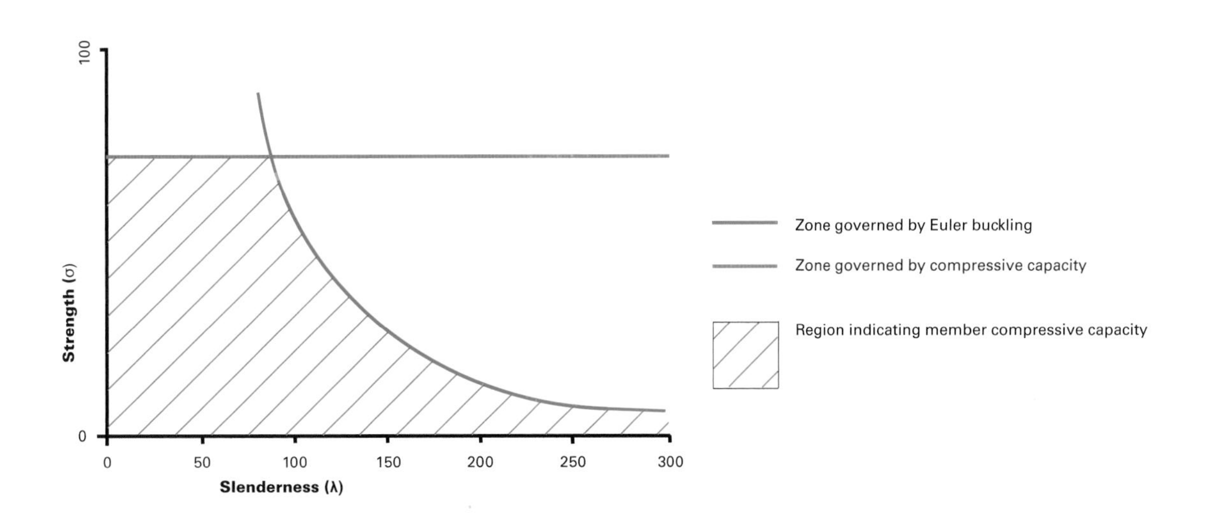

Effective lengths of columns with differing end restraints

L_o = Actual length
L_E = Effective length

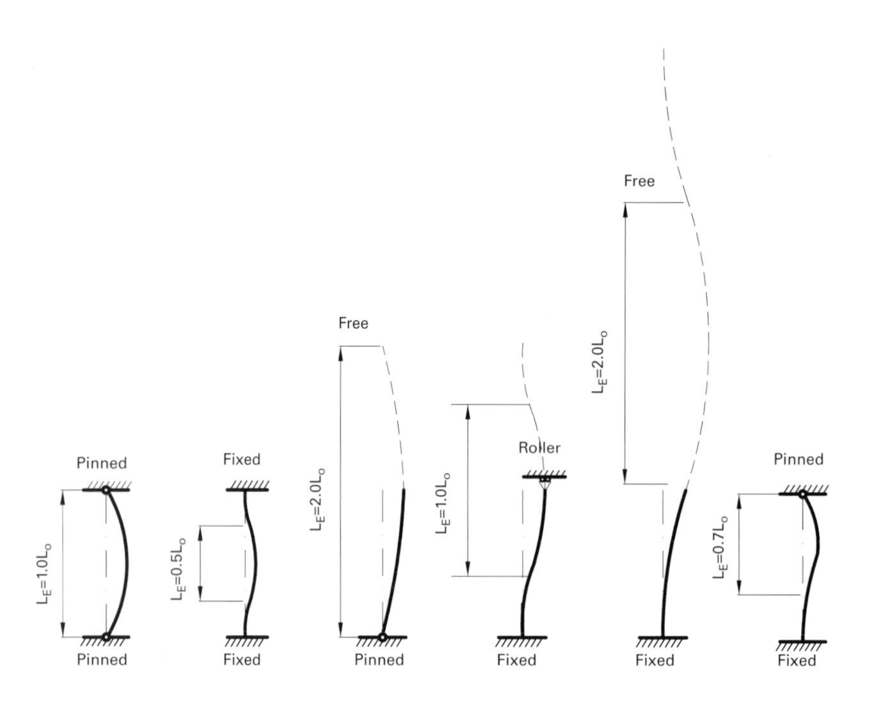

Standard universal beam and column sections

Universal beam section

Universal column section

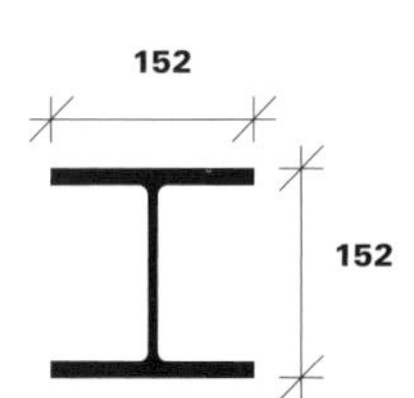

Section size: 152 x 152 UC 37 kg/m run

Second moment of area, I_x
and associated radius of gyration, r_x, in x axis:

I_x = 2210 cm^4
r_x = 68.5mm

Second moment of area, I_y
and associated radius of gyration, r_y, in y axis:

I_y = 706 cm^4
r_y = 38.7mm

Hence for this column section the ratio of the slenderness around the stronger x-axis against the weaker y-axis is 68.5/38.7 = 1.77

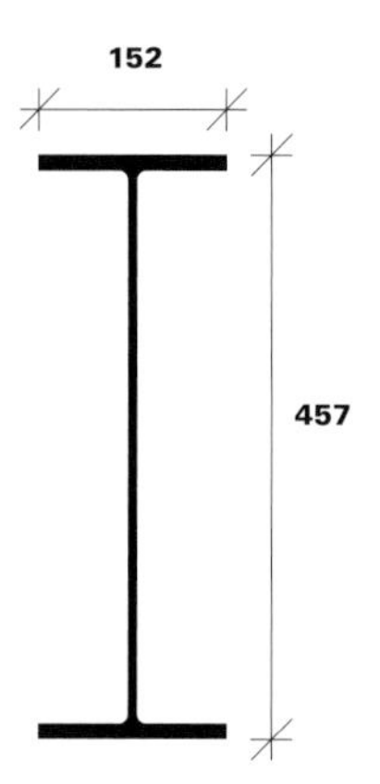

Section size: 457 x 152 UC 52 kg/m run

Second moment of area, I_x
and associated radius of gyration, r_x, in x axis:

I_x = 21400 cm^4
r_x = 179mm

Second moment of area, I_y
and associated radius of gyration, r_y, in y axis:

I_y = 645 cm^4
r_y = 31.1mm

Hence for this beam section the ratio of the slenderness around the stronger x-axis against the weaker y-axis is 179/31.1 = 5.76

2.1.5.3
Deflection

While 'bending moment' is a term for the internal force that is developed in a member under load, 'deflection' describes the extent to which a beam is displaced when loaded.

The formulae for calculating the deflection of beams under common loading and support conditions are indicated on the diagrams in section 2.1.3.

The second moment of area of a beam has a significant impact on the degree a beam will deflect, as can be seen from the equation below for a simply supported beam supporting a uniformly distributed load.

Deflection, $\delta = 5\omega L^4/(384EI)$

where
ω = applied load per metre
L = length of element
E = Young's modulus of material
I = second moment of area of rectangular cross-section ($= bd^3/12$)

As explained in section 2.1.5.1, the second moment of area of a rectangular section is directly related to the cube of the depth of the section. Therefore the deflection of a beam under uniformly distributed load is inversely related to the cube of the depth of the member. So increasing the depth of a member by a factor of 2 reduces the deflection of the system by 2 to the power 3, which is 8 times.

While it does not relate to section properties, it is worth noting that for a uniformly loaded element the deflection is also related to the length of an element to the power 4. Hence doubling the length of, for example, a 4-metre-long beam to 8 metres without changing its section properties will result in an increase in deflection of 2 to the power 4, or 16 times the original deflection. Increasing the span of a 4-metre beam to 5 metres without changing any of the section properties will result in the deflection increasing by nearly 2.5 times.

2.1.6
Fitness for purpose

So far, the performance of a structure has been examined in relation to how well it could resist applied loads without elements failing in terms of stress limits. This section examines another set of criteria that a structure has to meet to ensure that the building can serve the purposes for which it has been designed. These criteria are the 'serviceability' limit states, and generally relate to the movement of the structure under various loads.

2.1.6.1
Vertical deflection

A beam can be designed to be perfectly capable of resisting the stresses induced by an applied vertical design load, and therefore not pose any risk of causing a structural collapse, and yet still fail the serviceability deflection criteria, and hence be unsuitable.

Excessive vertical deflection of beams and slabs can cause the following problems:

- Perceived movement by building users, causing discomfort;
- Damage of finishes, such as ceilings and services, which may be supported by the deflecting structural members;
- Damage to the building cladding;
- Visually perceptible sagging of structural elements, causing concern/alarm.

The extent to which a structure can deflect vertically without exceeding any of the serviceability conditions is a function of the length of the span and the deflection under live load. Deflection under self weight is not relevant to the first three items in the list as this deflection would have already occurred prior to the application of the cladding, the services and the live loads, and hence would not be additional to any discomfort or damage that could occur. In order to limit the possibility of visual sagging, long-span beams can be fabricated with an upward curve that offsets some of the dead load deflection. This is called pre-cambering. Beams are often pre-cambered in the opposite direction to the deflection in order to cancel out the majority of the dead load deflection, thus reducing the overall perceived critical deflection.

The allowable deflection criteria vary slightly between materials and codes of practice, but in general the governing deflection criteria for beams and slabs are approximately as shown below:

Beam		Allowable dead + imposed deflection load	= L/200
	and	Allowable live load deflection	= L/360
	or	Allowable live load deflection for beams carrying brittle finishes (such as brick)	= L/500
Cantilever		Allowable dead + live load deflection	= L/100
	and	Allowable live load deflection	= L/180
		where L = length of beam	

In certain circumstances an increased deflection criteria is required. For example, in commercial buildings the cladding is often made from large glazed units that are susceptible to damage owing to the 'racking' effect if the supporting beam deflects significantly. To reduce this risk, edge members supporting large glazed cladding elements are often designed to meet L/1,000 or 12mm, whichever is the lesser of the two.

2.1.6.2
Lateral deflection

As with vertical deflection, the lateral deflection limits for most structures are determined to limit any perceived lateral movement and therefore discomfort to building users and limit damage to building elements. The maximum allowable deflection limit is related to the height of the building, and is often taken as: height/500.

2.1.6.3
Vibration

As well as in the case of deflection, a floor can also be deemed to fail serviceability requirements if it is subject to excessive vibrations. Vibrations can be caused by a single impulse force, such as the dropping of equipment, or, for example, an industrial process.

Essentially, vibration occurs when the floor structure oscillates. The amplitude and frequency of each oscillation will determine how perceptible the vibration is to the building user. Amplitude and frequency are functions of the span and stiffness of the floorplate, its self weight, the intrinsic damping within the floor and the force that is causing the vibration to occur. The assessment of a floor's vibration characteristics requires detailed calculations and is often undertaken using Finite Element software for all but the simplest of structures. Finite Element Analysis (FEA) is a method that can be used to create a mathematical model of a structure. The technique subdivides structural components into small pieces, or elements, and sets up mathematical equations that model the behaviour of and interaction between these elements, and thus the structure as a whole. These equations are then solved simultaneously in order to find an approximate solution; that is, to predict how the structure will behave as it is put under load. As the behaviour of each element affects and is affected by its neighbours, the calculations have to be repeated a number of times to take account of the effect of the neighbouring elements. This leads to a more accurate approximation of the behaviour. Further executions of the calculations will increase the accuracy of the analysis until a point when there is almost no difference between each subsequent repetition of the calculations. In FEA these calculations can be run many times, enabling very accurate models of components to be developed. Breaking the elements down into even smaller pieces further increases the accuracy of the FEA, but requires a greater number of calculations to be undertaken and therefore greater computing power.

The actual movement of a floor when subject to vibration is usually well within the allowable deflection criteria; however, the perception to a building user can be of much greater discomfort. The acceptable levels of vibration vary significantly between building usages, from industrial facilities at one end to laboratories and hospital surgeries at the other. A range of acceptable vibration criteria is available in design guides, advising on the maximum accelerations of the floor for different end-user conditions.

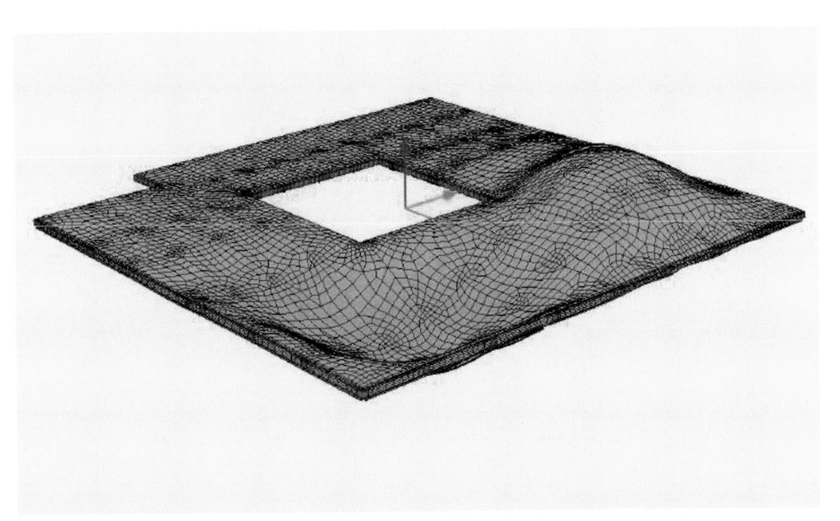

FEA computer analysis of floor vibration

2.1.7 Structures

2.1.7.1 Categories of structure

The previous sections have examined the loads applied to structural components, and how the material and sectional properties of those components contribute to their structural capabilities.

This section considers structures as whole entities of interconnected components and examines how they can be categorized and stabilized.

Heinrich Engel developed a system of categorization for structures, that was first published in 1965. He separated structural types into four categories:

- **Form active**
- **Vector active**
- **Surface active**
- **Section active**

These categories form a useful system for examining the primary structural drivers for the vast majority of structural forms.

Engel's categories provide designers with a useful framework within which structural forms can be grouped. Once the mechanism of load transfer in a building is identified, a designer can determine what parameters will and will not affect the structural efficiency of that building, and develop a design accordingly.

In reality, the practical requirements of achieving the required building function, form and aesthetic while supporting irregular loading conditions often determine that structural components will have to be designed to act in more than one mechanism of load transfer at any one time. For example, a floorplate may be designed to support vertical loads via a section active mechanism and simultaneously distribute lateral load to structural cores via a surface active mechanism. Similarly, arches and trusses are commonly required to support irregular loads that induce bending stresses in their components, thus reducing their structural effectiveness. In effect, most buildings are designed to compromise to some extent between pure structural efficiency and practical requirements.

Form active

'Form active' structures rely on a series of flexible, non-rigid components to achieve a stable form under loading. The most simplistic and easily apparent of these forms is a chain or rope bridge, which will deflect in order to reflect the position of any load placed upon it. Other, more three-dimensional examples include tensile fabric and gridshell structures, which when placed under tension also create stable forms that can be manipulated using double curves to create more interesting and more stable arrangements.

Pneumatic structures are further examples of structures whose form is directly related to the (hydrostatic) forces applied to them.

More common but less obvious examples of form active structures include arches. An arch can be considered to act in a similar manner to a loaded chain, with the exception that whereas the chain's components are in pure tension an arch's components are in pure compression.

In a perfectly efficient form active structure, the components are subjected to pure axial stresses (either compression or tension) only. If a point load is applied to the surface of a flexible form active structure, deformations will occur. Even rigid arches will develop bending under point loads (unless the load is applied vertically at the crown of the arch), which significantly reduces their structural efficiency.

1

2

3

4

5

6

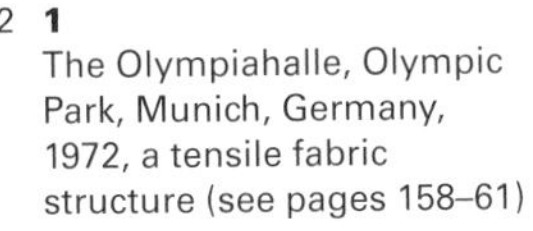

1
The Olympiahalle, Olympic Park, Munich, Germany, 1972, a tensile fabric structure (see pages 158–61)

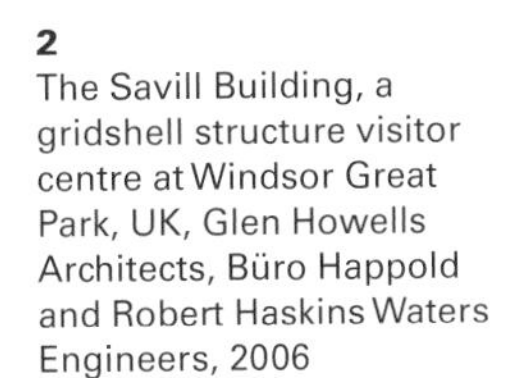

2
The Savill Building, a gridshell structure visitor centre at Windsor Great Park, UK, Glen Howells Architects, Büro Happold and Robert Haskins Waters Engineers, 2006

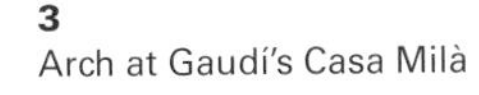

3
Arch at Gaudí's Casa Milà

4
Arch of the Winter Garden, Sheffield, UK, Pringle Richards Sharratt Architects and Buro Happold

5
Traditional stone arch of a Roman aqueduct, Segovia, Spain

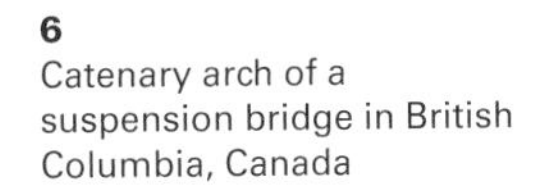

6
Catenary arch of a suspension bridge in British Columbia, Canada

Vector active

'Vector active' structural forms transfer load via a series of interlinked rigid components, which are small in comparison to the length of the overall structure and therefore not capable of developing significant bending or shear forces. The distribution of the externally applied force back to the points of support is subject to the directional and geometrical relationship between the components – hence the term 'vector active'. A simple two-dimensional truss is the most common example of a vector active structure. More complex examples include spaceframes and spherical or hemispherical dome structures.

The efficiency of a vector active structure is dependent on the individual members working in axial tension and compression only, rather than in bending. To achieve this the loads must be applied at or through the points where the members connect – known as the nodes. In reality it is often impossible to avoid some bending moment occurring in some members of a truss due to accidental loading scenarios or, in the case of bridges, due to the load of traffic on the bottom chord. Therefore, individual components of vector active structures are often designed with some additional sectional capacity to avoid instabilities developing.

1

2

3

4

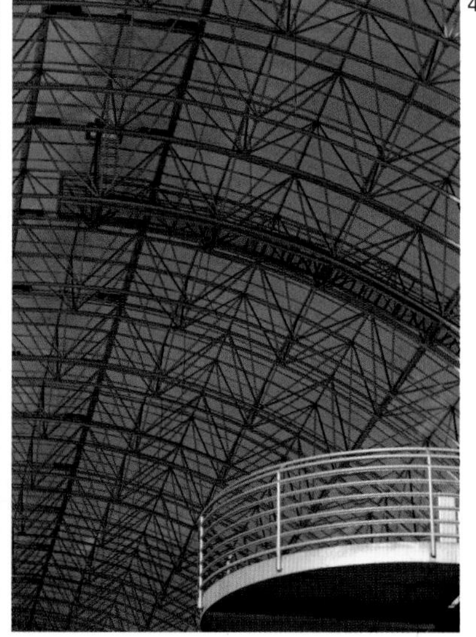

1
Cape Fear Memorial Truss Bridge, Wilmington, USA

2
Lamella Dome of the Palazzetto Dello Sport, Rome, Italy, Pier Luigi Nervi

3
Lamella Dome at Materials Park, South Russell, Ohio, USA, John Terrence Kelly

4
Detail of a spaceframe structure

Surface active

'Surface active' structures include concrete or masonry domes, cellular buildings and concrete shells. These are characterized by rigid surfaces that are capable of developing axial (compression and tension) and shear stresses. As with form active structures, any applied forces are redirected via the form or shape of the structures and therefore shape is intrinsically linked to structural performance.

The efficiency of a surface active structure is dependent on the form of the surface in relation to the forces applied to it. For example, the efficiency of a dome is driven by its height in relation to its span. A perfectly hemispherical dome is the most structurally efficient form in terms of material used and volume encapsulated.

Again, as with form active structures, surface active structures are poor at supporting point loads that generate local bending stresses. Openings within a stressed surface, or other discontinuities, also reduce the structural efficiency of the system.

When a surface active structure is designed purely to respond to the forces applied to it, it can be an extremely efficient form. For example, the reinforced-concrete roof to Smithfield Market in London forms an elliptical paraboloid that covers a column-free area of 68.6 x 38.1 metres and measures just 75 millimetres thick with a rise of 9.1 metres.

In many buildings the floor structure is designed as a horizontal surface active structural element, known as a diaphragm. This is used to transfer lateral loads into the vertically stiff elements of the building, such as shear walls or lift cores. This is expanded on in section 2.1.7.2.

1

2

3

1
Concrete Shell Aquarium, City of Arts and Sciences, Valencia, Spain, Santiago Calatrava and Félix Candela

2, 3
Interior and exterior views of the reinforced-concrete roof at Smithfield Market, London

Section active

'Section active' structures are the most versatile and most common form of structure in Engel's system. Section active structures rely on the sectional properties of individual rigid components, such as beams and columns, to support applied loads. All buildings that are constructed from beams, slabs and columns – from agricultural sheds to high-rise commercial buildings – can be described as section active. In contrast to form and vector active systems, the components of a section active system are designed to resist bending, shear and torsion forces as well as axial tension and compression.

The structural efficiency of a section active structure is dependent on the cross-sectional properties of the individual components and their unrestrained length and height.

1

2

3

1
Standard concrete structural frame

2
Standard steel structural frame

3
Standard timber structural frame

2.1.7.2
Stability

Any building, regardless of its particular load transfer mechanism, can be subjected to lateral forces. These are generated from either wind, seismic and/or 'out of tolerance' forces developed due to a lack of verticality in columns or the actual geometry of the building itself. In all cases the structure must be designed to be capable of transferring these lateral forces into the foundations. This must be done without overstressing any structural elements and without the building undergoing significant lateral deflections.

The extent to which a structure can be allowed to deflect under lateral loads is dependent on the use of the building and the material from which it is constructed. Typically, buildings are designed with an allowable lateral deflection limit of height/300 under the most onerous loading conditions such as a 1-in-50-year wind scenario. Buildings with brittle cladding, such as large glazed panels or brickwork, are more susceptible to damage and hence are often limited to overall height divided by 500. This is primarily to avoid damage occurring to the cladding elements rather than to avoid discomfort to the building's users, which will normally be negligible at such a level of deflection.

There are several fundamentally different methods by which a structure can be stabilized. The most common of these are explained in the following sections.

Tolerance of wind and seismic loads

1 Lateral wind load inducing lateral deflection of structure

2 Seismic ground movement inducing lateral deflection of structure

3 Geometry of structure induces lateral deflection of structure

i Initial profile of structure
ii Deflected profile of structure

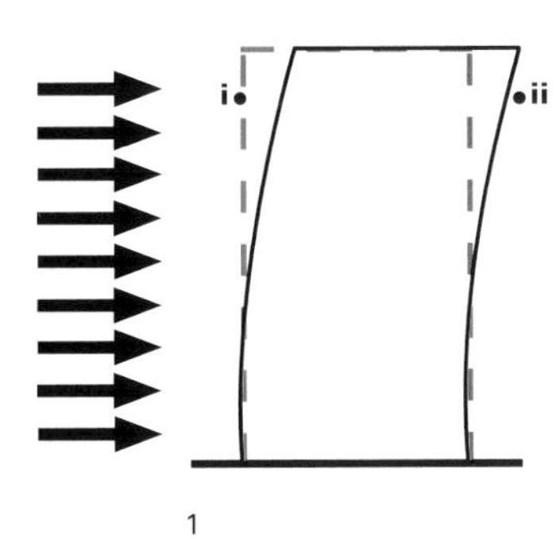

1

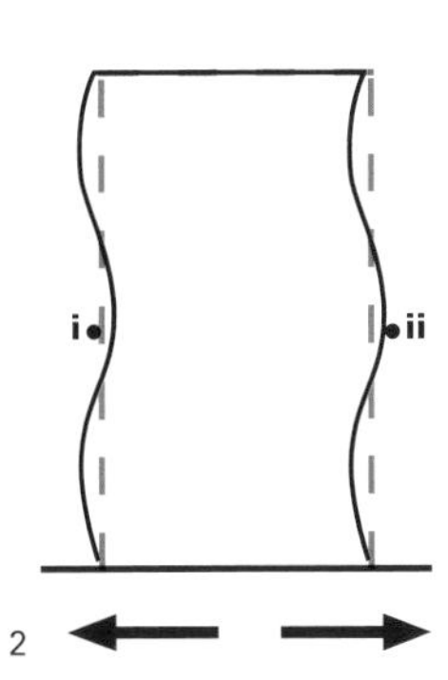

2

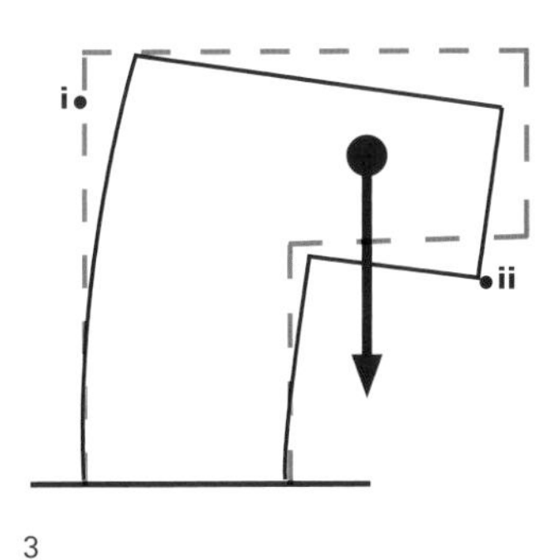

3

Rigid framed structures

Rigid framed structures are constructed from a series of columns and beams that form a frame onto which the building's cladding and floorplates are attached. They are typically constructed from steel, reinforced concrete or timber.

In rigid frames the resistance to horizontal loads is provided by a number of 'stiff frames' located throughout the structure. In each of these stiff frames the connection between the beam and the column is designed to be capable of transferring both the bending moment and the shear force that are developed by the applied horizontal forces (see diagram below). Since this stiff moment connection will not rotate, the frame will remain rigid under lateral load. The only lateral deflection that can occur will be due to the deflection of the vertical columns, which are designed to limit this deflection to within acceptable parameters. If the connections between the beams and columns of a frame are designed with pinned connections rather than moment connections, the frame would have no capacity to resist lateral loads and would form a mechanism that is by definition unstable.

The rigid frames within a multistorey building have to extend throughout the height of the building in order to be able to transfer the applied loads into the foundations. Any discontinuities caused by requirements such as double-height floors or locally removed columns will generate weak points and therefore necessitate larger, stiffer structural members in these locations to avoid exceeding the allowable deflection criteria.

Lateral loads, particularly wind, can be applied in all directions and so a rigid framed structure must be designed with frames orientated at right angles to one another to resist all possible loading scenarios.

The floor slabs that span between each frame in a rigid framed structure (and most other stabilizing systems) are often designed to act as diaphragms and distribute the lateral loads into each frame. In many cases the horizontal depth of the slab provides a sufficiently stiff element to ensure that it will not 'rack' when lateral loads are applied. Even a timber floor can be considered to be a stiff diaphragm when detailed correctly. The location of large openings in the floorplate must be carefully considered to ensure the diaphragm is not compromised.

Rigid frame under vertical and lateral loads

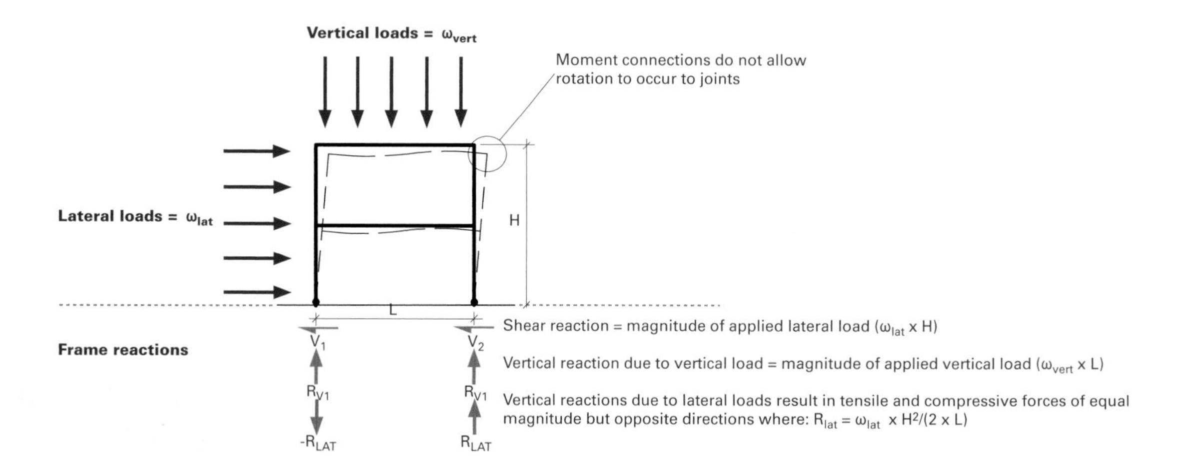

Pinned frame under lateral loads forming mechanism

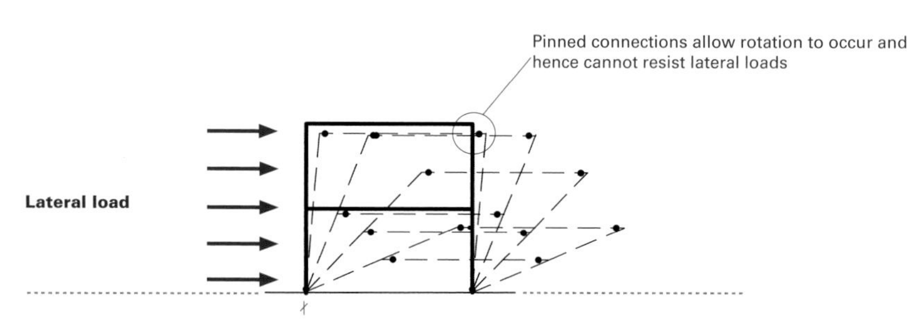

Rigid frames under lateral loads

Regular rigid frame under lateral load

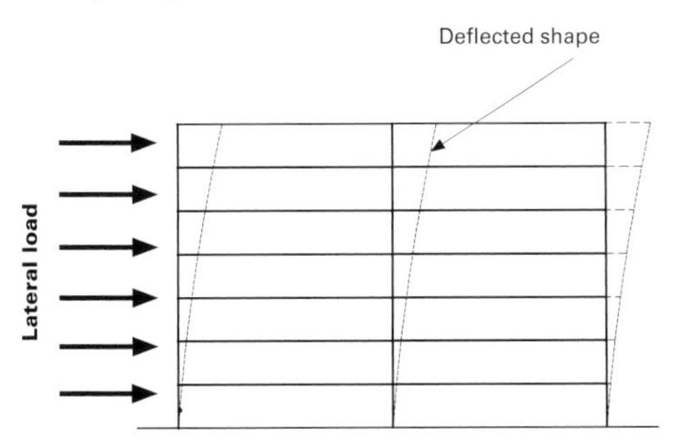

Rigid frame with increased floor span at 3rd floor level

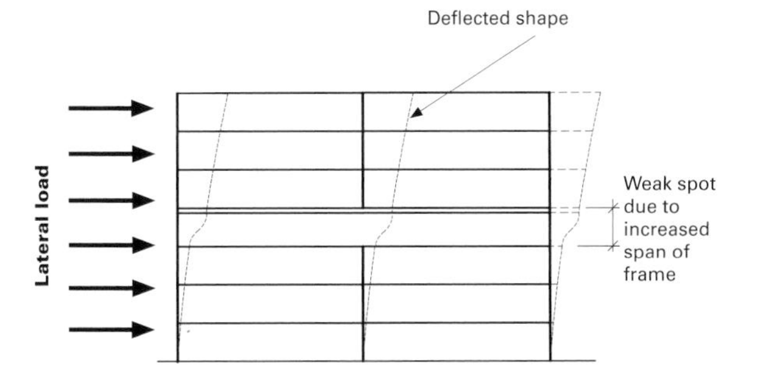

Rigid frame with double-height floor under lateral load

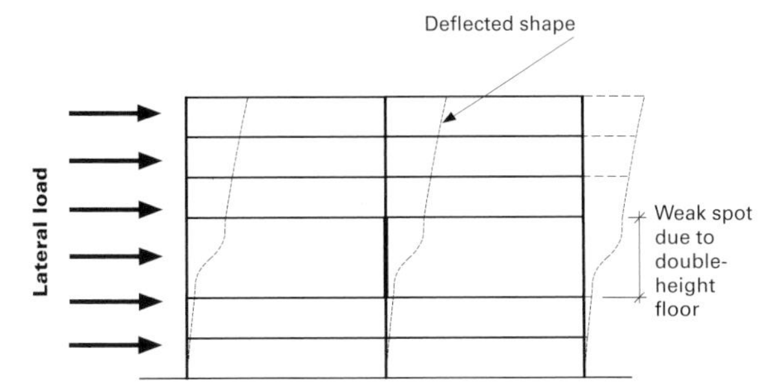

Floor slab acting as diaphragm

Rigid frame with stiff floor plate under lateral load

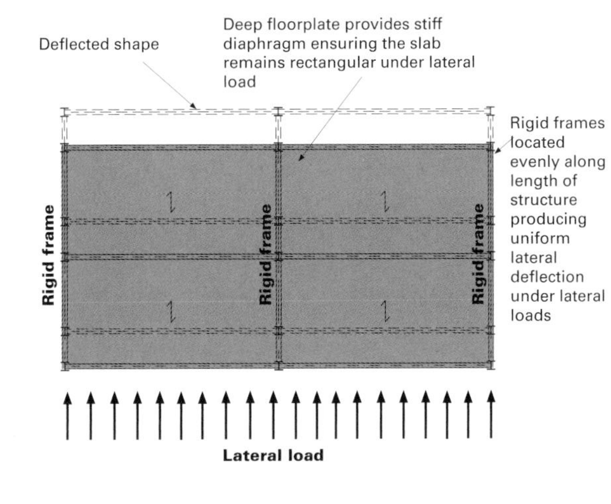

Rigid frame with floor plate containing significant openings under lateral load

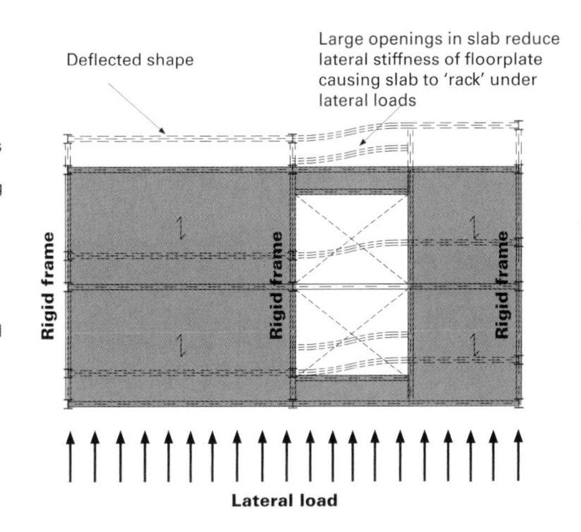

Braced framed structures

Braced framed structures, like rigid framed examples, are fabricated from a series of beams and columns linked together via a floorplate acting as a diaphragm. Unlike in rigid framed construction, in braced frames the beam-to-column connection is designed as a pinned connection, and thus is not capable of resisting applied lateral loads. Instead stability is provided via other elements such as shear walls, cores or braced frames located strategically throughout the plan of the building. These stiff elements – as is the case for the frames in a rigid framed structure – must continue for the full height of the building.

Ideally the location of the stiff cores, bracing or shear walls in a braced structure will be distributed evenly on plan. This will result in an even deflection of the building under lateral load. If the shear wall and cores are distributed non-symmetrically the structure can be subject to twisting under lateral loads.

Designing a braced structure can have the following implications in comparison to a rigid structure:

- Reduced cost of beam/column connections
- Reduced size and weight of beams and columns since they do not have to resist lateral forces
- Reduced complexity of beam/column connections make fabrication easier

The floorplans of rigid frame buildings, however, are not limited by the need for cores or shear walls and can therefore accommodate more open-plan arrangements than braced structures.

Such factors as the total height of a building, the height between each floor of a multistorey building, and the span between the columns all have a significant effect on frame stiffness. As frame stiffness reduces, column and beam sizes must increase to meet the deflection requirements. These factors significantly influence the efficiency of braced and rigid frames and can determine which is the most suitable option.

Rigid frame under lateral load

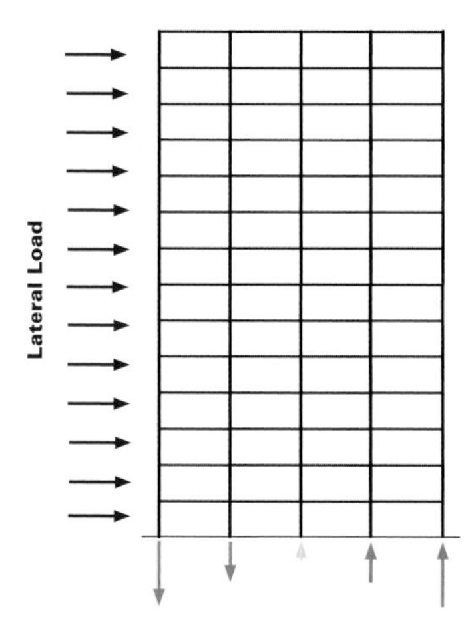

Bending moment due to lateral loads induces vertical tensile and compressive forces at the bases of the rigid frames

Braced frame with cross bracing under lateral load

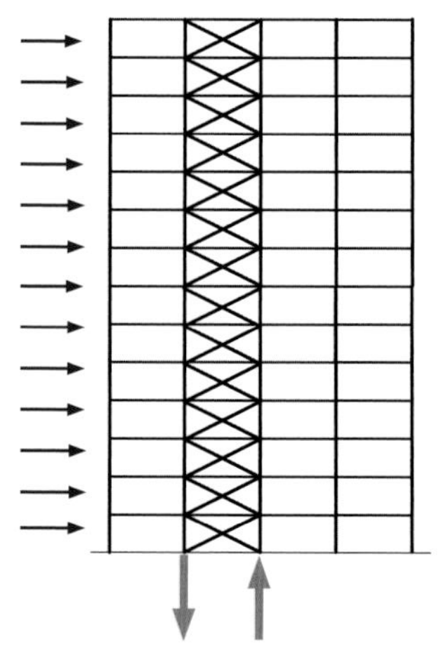

Bending moment due to lateral loads induces vertical tensile and compressive forces at the base of the cross-braced wall, adjacent columns support vertical loads only

Braced frame with shear wall under lateral load

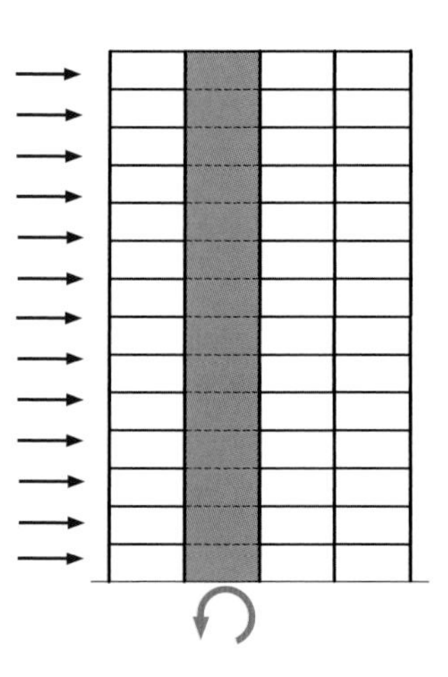

Bending moment due to lateral loads induces a bending moment at the base of the shear wall, adjacent columns support vertical loads only

Plan of braced structure with symmetrical stabilizing elements

i Centre of gravity of the building
ii Structural cores
iii Shear walls
iv Edge of building
v Extent of deflection under lateral loads

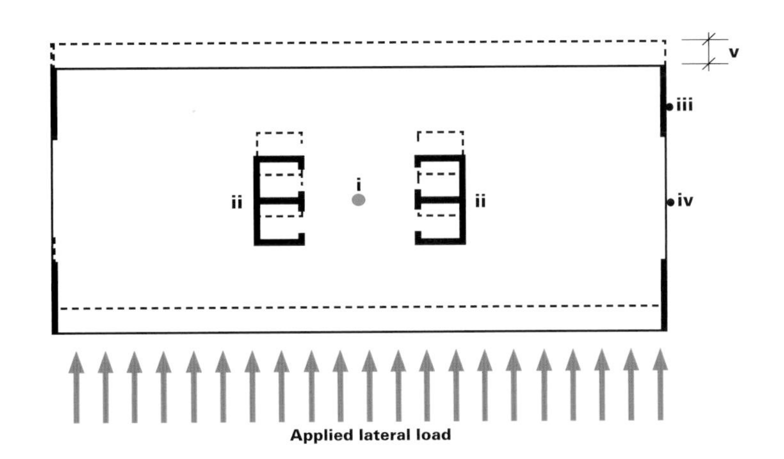

Plan of braced structure with non-symmetrical stabilizing elements. Nonsymmetrical arrangement of cores causes twisting of building under lateral loads

i Centre of gravity of the building
ii Structural cores
iii Shear walls
iv Edge of building
v Extent of deflection under lateral loads

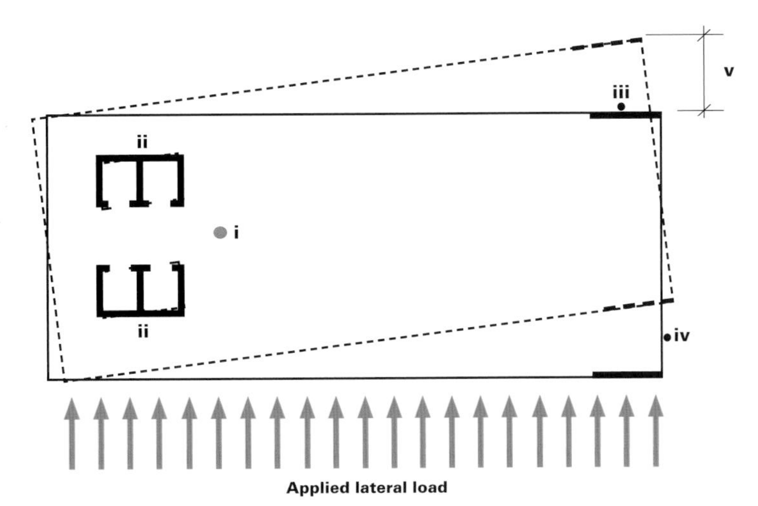

Concrete diaphragm with a structural core: The Shard under construction, London, Renzo Piano Building Workshop

Cellular structures

Buildings fabricated from a series of solid walls and floorplates can be described as cellular structures. The most common example of this form of construction is a typical masonry house with brick- and blockwork cavity walls and a timber joist floorplate with timber floorboards. Other examples of common cellular structures include *in situ* timber studwork structures, and pre-cast concrete and prefabricated steel buildings.

The stability of a cellular structure is provided by the walls, which act as surface active stiff panels that transfer the horizontal loads to the foundation level. Walls are significantly stiffer in their longitudinal axis than their transverse axis because stiffness is related to the cube of the depth of a member (see section 2.1.5.1). So the walls in a cellular structure must be distributed in both perpendicular directions to ensure the structure is capable of resisting the horizontal wind loads that can be applied in all directions.

As with framed buildings, the floorplates in a cellular structure have to act as diaphragms to transfer lateral loads into the stiff walls orientated parallel to the direction of the applied load. Large openings for staircases have to be located carefully to ensure that the floorplate remains stiff enough to distribute the forces effectively and to avoid distortion of the building under lateral load.

The floorplates and walls also have to be designed to be capable of resisting the vertical loads applied by the building's self weight and its occupants. As mentioned previously, this is an example where a structural element is designed with two distinct load transfer mechanisms: surface active to transfer the horizontal loads and section active to support the vertical loads.

In summary, for cellular systems to be effective the following conditions have to be achieved:

i) Structural wall panels are vertically continuous through the height of the building, thus providing a direct stress path for loads to the foundations and avoiding transfer structures.
ii) The floorplate must be capable of acting as a diaphragm. Timber joists in particular require blocking and to be positively fixed to the floorboards.
iii) Holes in the floor should be located to allow sufficient connectivity between the floorplate and the stabilizing walls.
iv) The floorplate must be positively connected to the stabilizing walls to enable the shear forces to be adequately transferred.

Typical cellular building plan – coloured walls indicate the elements providing effective lateral restraint under each lateral load condition

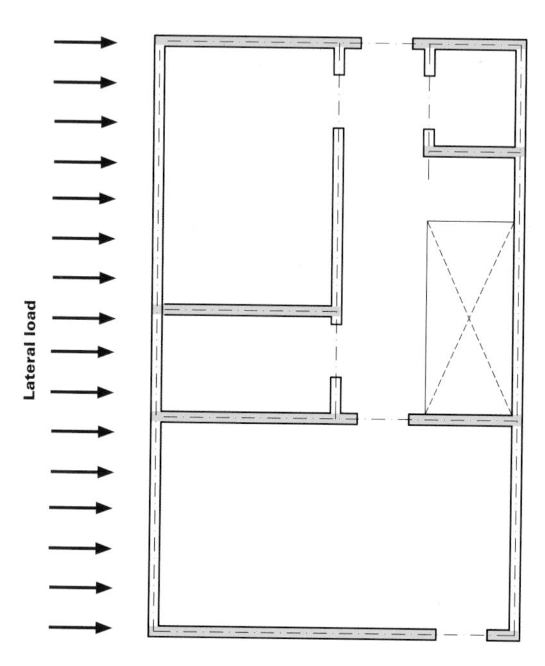

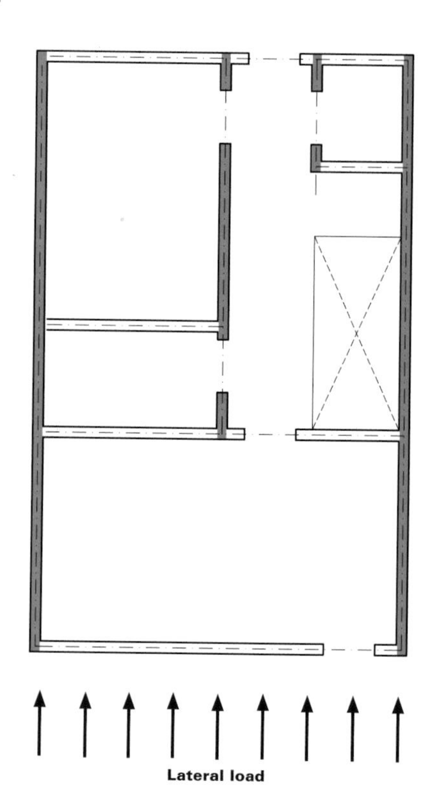

Structures inherently resistant to lateral forces

Certain structures are inherently stable owing to their form. These include many form and surface active examples such as domes, shells, gridshells, cable net and tension fabric structures. All of these can be designed to resist lateral forces without the need for any additional stabilizing elements. The stability system of domes and tension fabric structures are examined on the following sketches.

Tension fabric structure

T = tension
C = compression

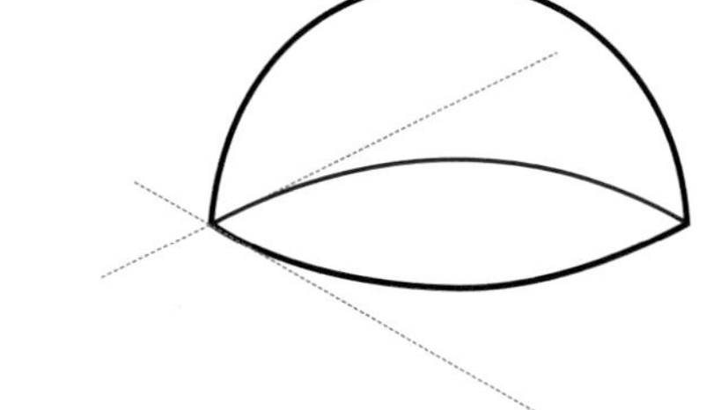

Dome structure

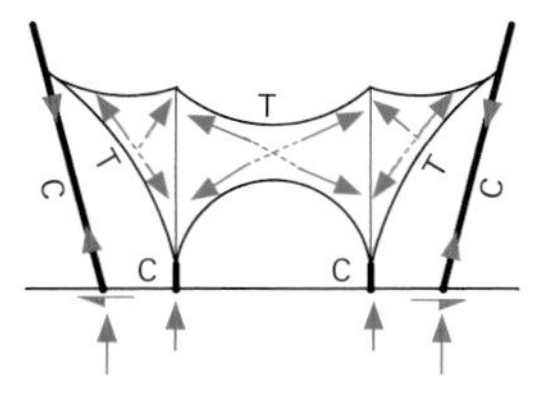

Force diagram under self weight only

Fabric is stretched over steel frame in a double-curved, three-dimensional form, which induces tensile forces in the fabric and compressive forces in the frame.

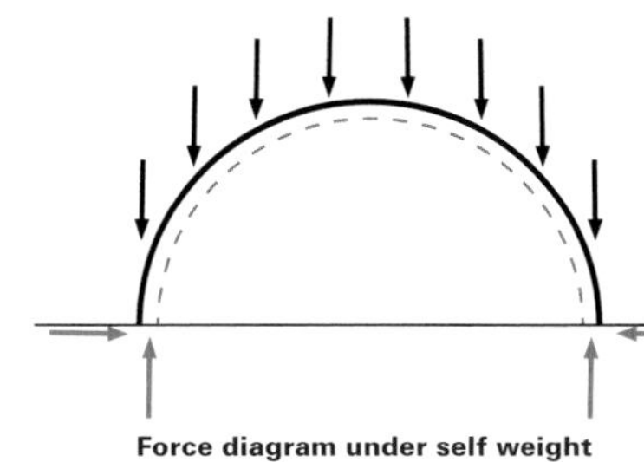

Force diagram under self weight only

Vertical load induces a horizontal thrust through the arched structure. Magnitude of horizontal thrust is dependent on the weight and profile of the dome.

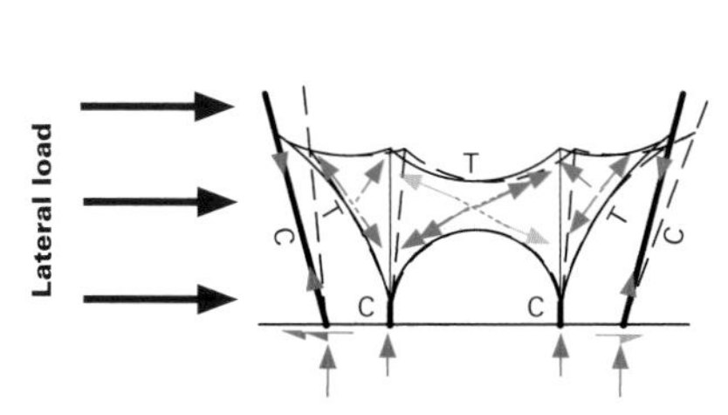

Force diagram under self weight and lateral loads

Under lateral load the tension in the fabric in one direction increases as the lateral forces are transferred through it into the supporting steel frames.

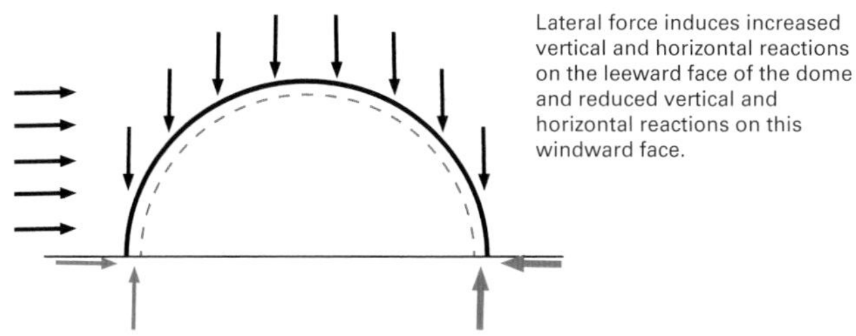

Force diagram under self weight and lateral load

Lateral force induces increased vertical and horizontal reactions on the leeward face of the dome and reduced vertical and horizontal reactions on this windward face.

Structures such as igloos **1**, airship hangars **2** and tensile fabric structures, for example London's Millennium Dome **3**, can resist lateral forces without any additional stabilizers

1

2

3

2.1.7.3
Towers

As buildings extend in height, structural stability becomes increasingly difficult to achieve owing to the relative reduction in the height-to-width ratio. When a structure approaches 30 storeys or, say, 120 metres high, alternative, more complex stability systems are needed to provide resistance to lateral forces. These different systems can be separated into two distinct groups:

- **Interior structures**
- **Exterior structures**

Interior structures are so called because their stability system is essentially located within the interior of the building via cores or shear walls. Exterior structures, on the other hand, use the perimeter skin of the building to form a stiff tube to provide stability.

Examples of each of the subgroups of these typologies, together with the ranges over which they are efficient, are provided in the following tables.

Towers – interior structures

Frame type	Frame description	Approx. efficient number of storeys	Example building		
Rigid or braced frame	Either structural cores, shear walls or rigid beam-to-column connections provide lateral stability	>30	One Canada Square, London 50 storeys 235m high		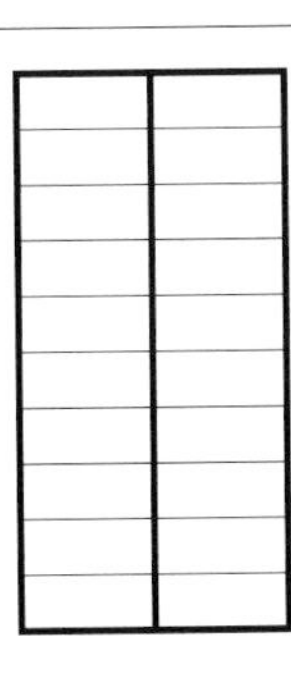
Outrigger construction	Cores provide stability, with additional stiffness afforded via 'belt trusses' spaced at regular storey intervals, which are linked to outrigger columns; this arrangement increases the lever arm over which the lateral loads are distributed	>100	Taipei 101, Taiwan 101 storeys 509m high		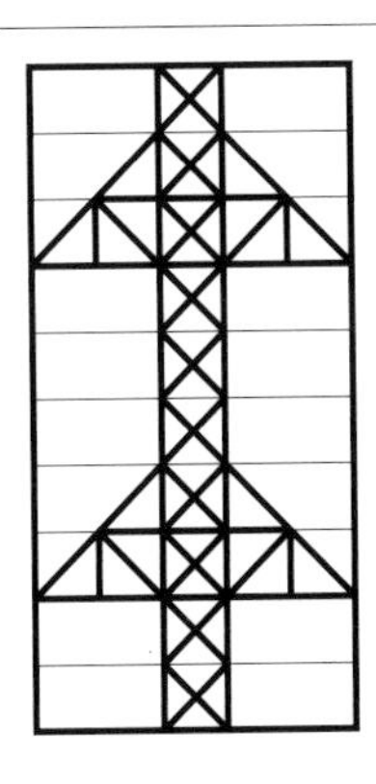

Towers – exterior structures

Frame type	Frame description	Approx. efficient number of storeys	Example building		
Framed tube	A series of closely spaced perimeter columns are connected to the floor via fixed connections, which allows them to act as a single large external core. This has a greater width than an internal core, making it more efficient.	100+	Aon Center, Chicago 83 storeys 346m high		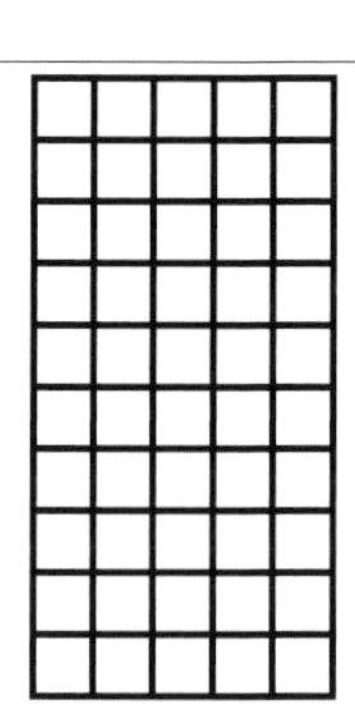
Braced tube	A similar system to the framed tube, but the stiff outer core is formed via a braced truss rather than closely spaced columns	100+	John Hancock Center, Chicago 100 storeys 344m high		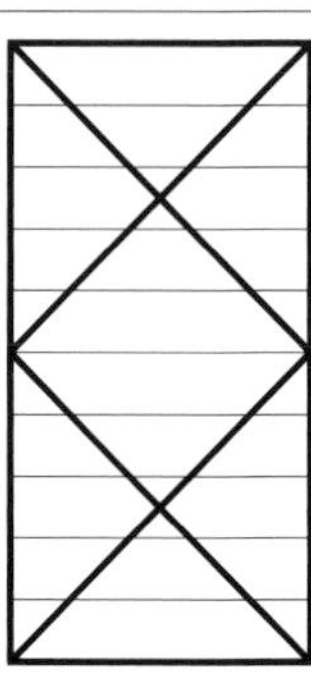
Bundled tube	A series of framed tubes bundled together to utilize the increased strength of several tubes structurally linked. As the building height increases, some tubes are terminated	100+	Willis (formerly Sears) Tower, Chicago 108 storeys 442m high		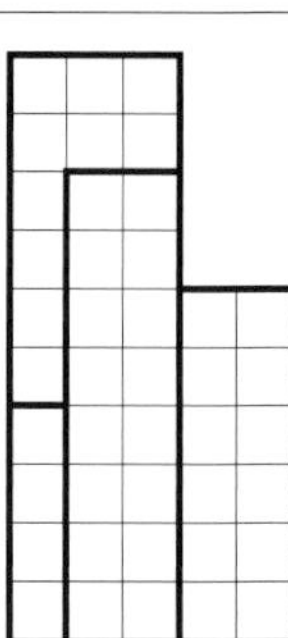
Diagrid	Uses a 'diagrid' rather than a braced truss to form the stiff perimeter tube	<100	Swiss Re Tower, London 41 storeys 181m high		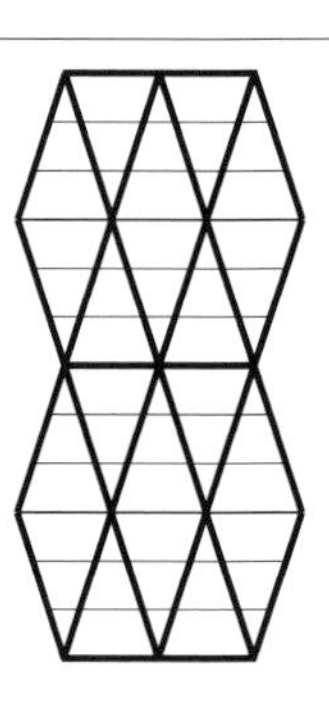

2.2
Structural systems

2.2.1
Introduction

This section examines the most common structural-frame materials used in building construction: steel, reinforced concrete and timber.

Section 2.2.2 discusses the attributes of each material against a series of criteria that help determine the suitability of each structural material in specific building types.

Section 2.2.3 then provides general data for structural components including rules of thumb and economical span ranges.

2.2.2 Structural material assessments

Structural steel construction assessment

Common usage in UK	All sectors, particularly high-rise
Structural performance	The inherent strength of steel determines that it is capable of spanning relatively long distances efficiently. This enables buildings to have larger grids with fewer columns
Weight	In general steel-framed buildings weigh less than concrete-framed ones, and therefore exert smaller loads on to their foundation system
Deflection	Deflection, as opposed to stress failure, is often a critical design criterion for steel beams – particularly long-span beams. This can be limited by pre-cambering up to two-thirds of the dead load applied to a steel beam
Vibration	As steel frames are often lightweight and relatively long-span, they can be susceptible to adverse in-use vibrations. This must be identified and designed out at design stage – by reducing spans, increasing permanent dead loads or stiffening the system
Fire protection	Steel has virtually no inherent fire resistance and normally requires additional measures, such as sprayed or painted coatings applied directly to its surface or boarding with fire-resistant material, to achieve the necessary fire protection
Programme	Steel frames can be erected very quickly in comparison to concrete frames, reducing construction programmes. However, the use of following trades such as cladding and fire protection can offset this programme advantage
Sustainability	The environmental performance of a steel-framed building in comparison to a concrete-framed building is subject to many variables, and should be examined on a case-by-case basis
Cost	The cost of a steel frame is generally driven by the weight of the steel used
Flexibility	Steel frames can be relatively easily strengthened or adapted post-construction

Reinforced concrete construction assessment

Common usage in UK	All sectors
Structural performance	Reinforced concrete can be designed to span long or short distances depending on the depth and volume of reinforcement used. Post-tensioning of the concrete can be used to further increase the distances that can be efficiently spanned
Weight	In general, concrete-framed buildings weigh more than steel-framed buildings and therefore exert larger loads onto their foundation system
Deflection	The deflection of concrete elements is normally governed by the depth of the beam in relation to its span. Cambering of the formwork can be used to reduce the dead-weight deflections
Vibration	The heavy nature of long-span concrete frames reduces the risk of problems due to vibration; however, this must still be examined at design stage
Fire protection	Concrete has excellent inherent fire protection, achieved via the 'cover' it affords to the reinforcing bars. Cover can be increased to achieve higher protection as required
Programme	*In situ* concrete frames take longer to construct than steel frames. Pre-cast frames can be constructed in a similar timescale as that for steel frames. Overall building programmes can be reduced if exposed concrete finishes are used, as this can negate the requirement for following trades for suspended ceilings and some cladding
Sustainability	The environmental performance of a concrete-framed building in comparison to a steel-framed building is subject to many variables, and should be examined on a case-by-case basis. The inherent thermal mass of a concrete frame can be used in the environmental strategy of a building; however, the cooling/heating design strategy must be developed to utilize this to achieve the maximum benefit
Cost	The cost of a concrete frame is generally driven by the concrete volume, the mix design, the reinforcement volume and the formwork type
Flexibility	*In situ* concrete is 'poured' and can therefore be formed into any shape more easily than other materials. Existing concrete frames, in general terms, are more difficult to adapt than steel frames as the reinforcement is not visible and hence any post-construction redesign is reliant on the existence and accuracy of record drawings or intrusive examination. Post-tensioned concrete is complicated further owing to the requirement to avoid damage occurring to the post-tensioned tendons

Timber construction assessment

Common usage in UK	Residential, education. Typically low-rise (up to 5 storeys)
Structural performance	Timber frames generally are designed to span shorter distances than concrete or steel. Glue-laminated timber and Laminated Veneer Lumber (LVL) are products that have been manufactured to increase the structural performance of timber. The grade of timber has a large bearing on its ability to resist load
Weight	The lightweight nature of timber makes it an excellent material for long-span lightly loaded roof structures or pedestrian bridges
Deflection	As with other materials, deflection is related to section depth
Vibration	As timber is primarily used to span shorter distances than other structural materials, vibration is often not a critical design driver
Fire protection	Requires significant fire protection
Programme	Prefabricated timber frames with pre-applied insulation can facilitate a very rapid construction programme
Sustainability	Arguably the only truly renewable construction material, if sourced responsibly. Overall environmental performance of the structure is still subject to many factors and should be examined on a case-by-case basis
Cost	Generally low-cost material, but the selection of connections (mechanical) can have a significant effect on cost
Flexibility	Highly flexible and adaptable

2.2.3
Structural components

As with any 'rule of thumb' system, the information provided in this section is of a general nature and is sufficiently accurate to provide a basic feel of the section sizes and depths required at the early design stage, but should always be proved via detailed calculations as the design progresses.

The information in this section is divided into coverage of the structural elements – including beams, slabs and columns – and then subdivided according to the various materials commonly used to form these elements.

2.2.3.1
Beam systems

Properties of steel beams

The aspect ratio between primary and secondary beams can significantly affect the load a system can support, and hence will impact on the performance of the beams. The rules of thumb in this table assume that an aspect ratio of approximately 1:3 is achieved.

Beam type	Comments	Typical span range	Typical span/depth ratio
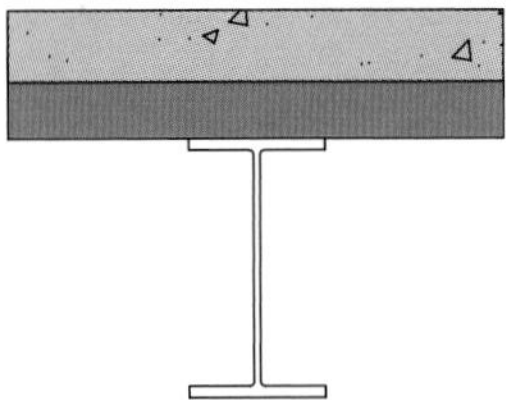 **Standard rolled universal beam (UB) beam section**	**1** Commonly used with pre-cast concrete, concrete-infilled metal decking and timber joist floors **2** Services pass either below or above structural beam **3** Shelf angles can be used with pre-cast planks to reduce structural depth of floor	Secondary beams 10–15m Primary beams 6–10m	15:1
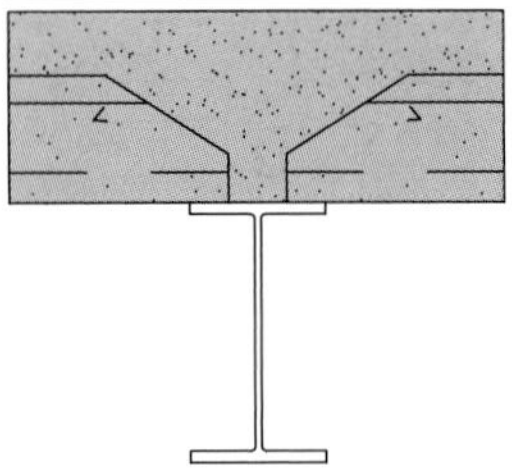 **Standard steel beam with pre-cast slab and structural toppings**	**1** Commonly used with pre-cast concrete, concrete-infilled metal decking and timber joist floors **2** Services pass either below or above structural beam **3** Shelf angles can be used with pre-cast planks to reduce structural depth of floor	Secondary beams 10–15m Primary beams 6–10m	15:1
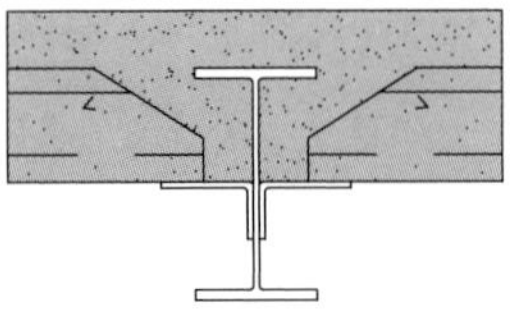 **Standard steel beam with pre-cast slab on rolled steel angles**	**1** Commonly used with pre-cast concrete, concrete-infilled metal decking and timber joist floors **2** Services pass either below or above structural beam **3** Shelf angles can be used with pre-cast planks to reduce structural depth of floor	Secondary beams 10–15m Primary beams 6–10m	15:1
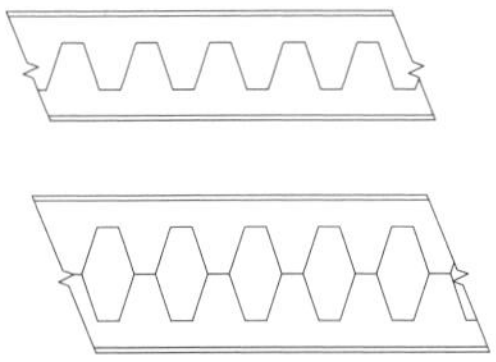 **Castellated rolled steel beam**	**1** Commonly used in floor construction with pre-cast or concrete-infilled metal decking **2** Castellations are formed by profile-cutting the web of standard rolled sections such that when the upper section is lifted and moved laterally an octagon is formed in the web between the upper and lower sections. The web is then re-welded, creating a deeper section **3** Services can be designed to pass through the octagonal openings **4** At supports and points of high point-loading, the octagonal holes are often infilled with steel plate to increase shear strength **5** Holes measured typically 0.6 x D in width and are at 0.75 x D centres, where D is the depth of the castellated beam	14–17m	18:1
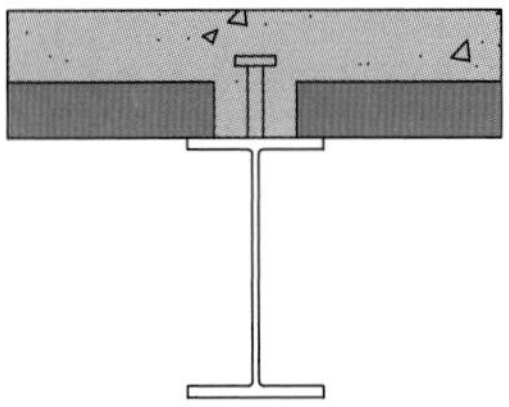 **Composite rolled steel beam with concrete slab on metal decking**	**1** Commonly used with pre-cast concrete planks or concrete-infilled metal decking **2** Shear studs welded to the top flange of the steel provide a shear key between steel and concrete – allowing the concrete and steel to act compositely with the concrete slab, effectively forming a wide top flange to the steel beam	Secondary beams 10–18m Primary beams 7–11m	20:1 steel-beam depth only

Properties of steel beams

The aspect ratio between primary and secondary beams can significantly affect the load a system can support, and hence will impact on the performance of the beams. The rules of thumb in this table assume that an aspect ratio of approximately 1:3 is achieved.

Beam type	Comments	Typical span range	Typical span/depth ratio
Asymmetrical steel beam (ASB)	**1** Commonly used with pre-cast concrete planks or concrete-infilled metal decking **2** The wider bottom flange allows the slab (pre-cast or metal decking) to be constructed within the depth of the steel beam, thus reducing the overall construction depth **3** ASB sections are generally more expensive than standard rolled sections, but can provide significant floor-depth savings	5–9m	25:1 steel beam depth only
Composite fabricated steel beam with concrete slab on metal decking	**1** Used when standard rolled beams are inappropriate owing to insufficient strength, stiffness or limited construction depth **2** Fabricated via steel plates welded together **3** Used particularly to achieve long spans with shallow construction depths **4** Typically used with pre-cast planks or concrete-infilled metal decking **5** Cellular holes are often cut into the fabricated beam to reduce weight and allow services to pass through the beam, thus reducing overall construction depth	10–18m	22:1 steel beam only
Steel truss	**1** Used for long-span roofs and more heavily loaded floor structures **2** Can be designed as composite or non-composite **3** All truss connections are pinned **4** There are many different forms of truss that have been designed all over the world, particularly in bridge construction. These include: Allan truss Bailey bridge truss Bollman truss Bowstring arch truss Brown truss Lenticular truss Burr arch truss Cantilevered truss Fink truss Howe truss King post truss Queen post truss K truss Lattice truss Warren girder truss	15–90m	10:1 can vary considerably
Vierendeel truss	**1** A truss with no internal diagonal elements, in which all of the connections are fixed-moment connections **2** Less structurally efficient than standard trusses with diagonal members, but the omission of diagonal members allows clear path for services and/or building users to pass	15–45m	10:1
Trussed steel arch	**1** Commonly used as long-span roof structures, such as train stations or bridges **2** Arches rely on lateral restraints to resist spreading forces generated within the arch structure. This can be achieved via a lateral restraint at the support or by tying the base of the arch together with a tension member known as a bowstring	From 25m upwards St. Pancras Station, London, spans 73m. Trussed-arch bridges span in excess of 500m	5:1

Properties of steel beams

The aspect ratio between primary and secondary beams can significantly affect the load a system can support, and hence will impact on the performance of the beams. The rules of thumb in this table assume that an aspect ratio of approximately 1:3 is achieved.

	Beam type	Comments	Typical span range	Typical span/depth ratio
	Tied steel arch Sydney Harbour Bridge	**1** Commonly used as long-span roof structures, such as train stations or bridges **2** Arches rely on lateral restraints to resist spreading forces generated within the arch structure. This can be achieved via a lateral restraint at the support or by tying the base of the arch together with a tension member known as a bowstring	From 25m upwards	5:1
	**Steel space frame** Montreal Expo Dome	**1** Typically used for long-span lightweight roof structures with limited points of support **2** Frames span in multiple directions as opposed to unidirectional truss structures, making spaceframes extremely efficient **2** All connections are pinned	5–60m but can go up to 150m	22:1
	**Steel domes (geodesic)** Eden Project	**1** Typically used for long-span lightweight roof structures in stadia or theatre spaces **2** Structurally similar to spaceframes but curved in two directions. Several different variations of dome have been developed, including the Schwedler, lamella, lattice and geodesic types **3** All connections are pinned	Up to 85m	22:1
	**Steel domes (lamella)** Louisiana Superdome	**1** Typically used for long-span lightweight roof structures in stadia or theatre spaces **2** Structurally similar to spaceframes but curved in two directions. Several different variations of dome have been developed, including the Schwedler, lamella, lattice and geodesic types **3** All connections are pinned	Up to 85m	22:1
	**Steel portal frames**	**1** Typically used for long-span single-storey structures, such as warehouses or agricultural buildings, where haunches have little impact on ceilings or services co-ordination **2** Portal frames account for approximately 50 per cent of the structural-steel usage in the UK	20–60m	N/A Spans driven by haunch size

Properties of concrete beams

Beam type	Comments	Typical span range	Typical span/depth ratio
Reinforced concrete beam	**1** Commonly used with *in situ* concrete slabs. See section 2.2.3.2 for details of possible concrete slab arrangements **2** At initial design stages, beam width can be assumed as: depth/2.5 **3** Beams can be designed as rectangular (typically), or as flanged beams if the slab on either side of the beam is continuous for the entire length of the beam **4** Services typically pass above or below concrete beams **5** A differentiation is made for concrete structures as to whether the beams are continuous or simply supported. (See section 2.1.7.2 'Rigid Frames' for a description of a continuous or moment connection.) Generally in multi-span bays it is likely that the beams will be continuous, while in single-span bays they are likely to be simply supported	6–10m	Continuous beam 26:1 Simply-supported beam 20:1 Cantilevered beam 7:1
Post-tensioned reinforced-concrete beam	**1** Post-tensioning is a specialized construction process involving a series of high-tensile steel tendons being cast within a concrete beam and then tensioned up as the concrete beam begins to cure. Owing to a predetermined curve in the tendons, this tensioning induces a compression force into the soffit of the beam and a tensile force on the upper face. These pre-induced compressive and tensile stresses offset some of the stresses induced in the beam element as it is loaded **2** Post-tensioned concrete beams can span longer distances than traditional reinforced-concrete beams **3** Slab depths are reduced, leading to less material and therefore less load owing to self weight **4** Post-tensioned beams are usually used in conjunction with post-tensioned concrete slabs, and with wide beams measuring approximately: span/5	6–15m	22:1
Pre-cast prestressed reinforced-concrete beam	**1** Shallow (150–225 millimetres deep) pre-cast prestressed beams in conjunction with infill blocks are often used in residential floor construction **2** Full pre-cast prestressed concrete frames are typically used in commercial developments where a fast construction programme is required **3** Deep (750–2,000 millimetres deep) pre-cast prestressed beams are typically used in bridge construction where a fast construction programme is required **4** Connections between pre-cast beams and supporting columns require careful detailing	Pre-cast prestressed floor beams 4–6m Pre-cast prestressed frames 5–11m Pre-cast prestressed bridge beams 10–50m	22:1 Member depth is often influenced by detailing

Properties of timber beams

Beam type	Comments	Typical span range	Typical span/depth ratio
Standard sawn-timber beams	**1** Typically used in residential and light commercial developments **2** Strength subject to grade of wood **3** Require fire protection usually provided via plasterboard ceilings **4** When sourced correctly, is the most sustainable of structural materials **5** Larger sections of up to 300 x 300 millimetres, known as 'bressemer' beams, can be sourced. Range and span-to-depth ratios relate to typical timber joists measuring up to 300 millimetres deep by 50 millimetres wide	3–6m	20:1 Subject to grade of timber, width and spacing of joists

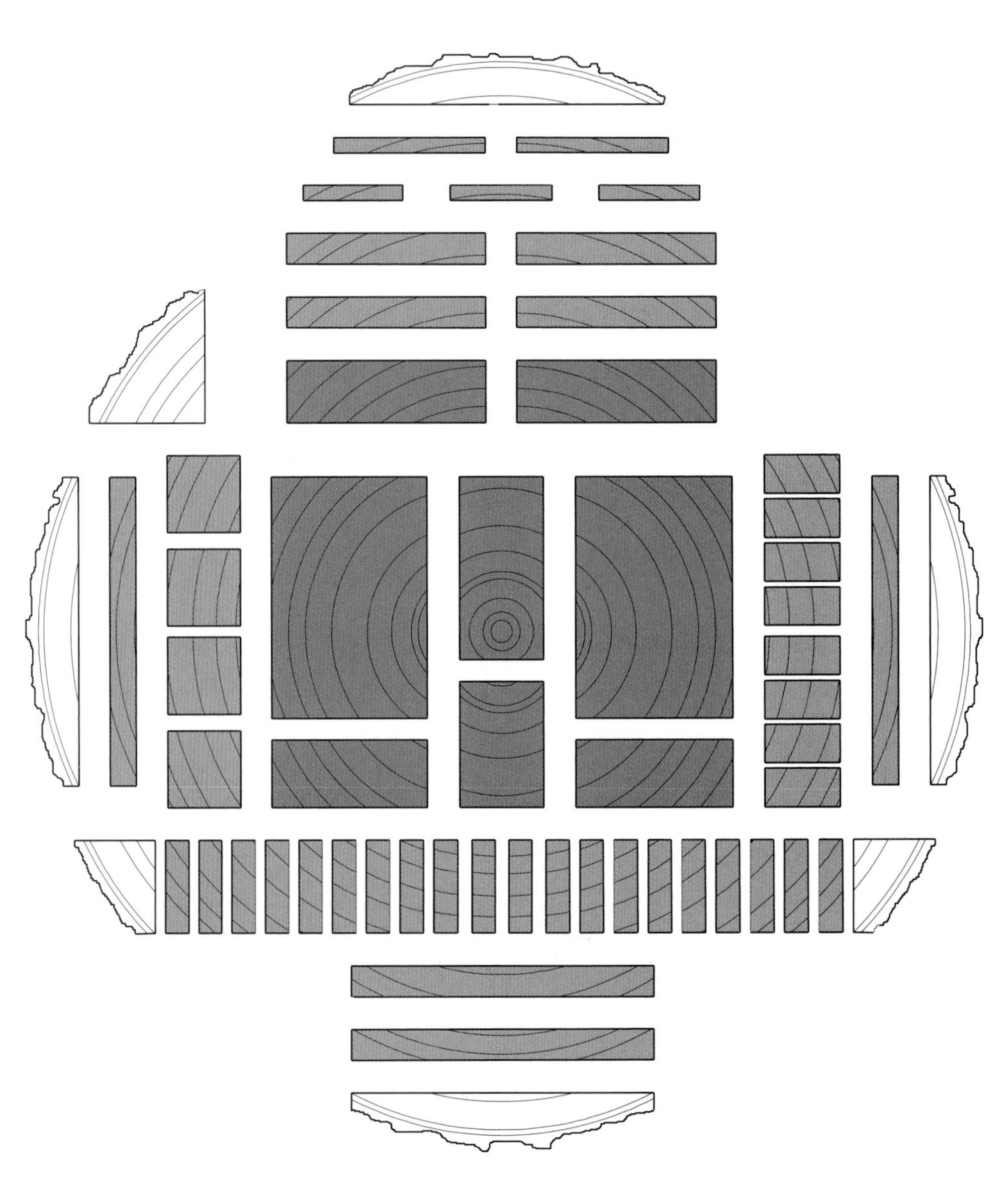

Typical structural timber cuts from a tree cross-section

Properties of engineered timber products

Engineered timber products include Glue-laminated beams (glulam), Laminated Veneer Lumber (LVL) and Laminated Strand Lumber (LSL).

Each of these products is fabricated from layers of sawn timber, which are glued together to form the beam. This process increases the homogeneity of the final product as all the imperfections within sawn timber, such as knots, are distributed along the beam rather than being concentrated at particular positions. This in turn increases the strength of the element.

The fabrication process also reduces the tendency of the members to warp, twist or bow.

Engineered timber products can be fabricated to a range of section sizes and lengths.

Beam type	Comments	Typical span range	Typical span/depth ratio
Glulam beams	**1** Typically used in lightweight timber roofs (often exposed) or light commercial structures **2** Can be fabricated to significantly longer lengths than standard sawn-timber joists **3** Strength subject to grade of timber used and number of laminations, and is advised by specific manufacturer	Roof beams 6–20m for standard section sizes Can increase to <50m with non-standard sizes Floor beams 4 –14m for standard section sizes	20:1
Laminated Veneer Lumber (LVL) and Laminated Strand Lumber (LSL)	**1** Manufactured in panels measuring approximately 16.5 x 2.95 metres and up to 300 millimetres thick. Can be used as simple beams similar to glulam, but more commonly used in panel form to create structural building forms **2** Typically used in residential, educational or light commercial structures **3** Can be fabricated to significantly longer lengths than standard sawn-timber joist **4** Strength subject to grade of timber used and number of laminations, and is advised by specific manufacturer **5** Ranges and span-to-depth ratios similar to glulam.	Similar to glulam	20:1
Timber I-sections	**1** Manufactured with either sawn-timber or LVL flanges and a plywood or Oriented Strand Board (OSB) web **2** Typically used in residential or light commercial structures **3** Can be fabricated to significantly longer lengths than standard sawn-timber joists **4** Strength subject to grade of timber used and number of laminations, and is advised by specific manufacturer **5** Ranges and span-to-depth ratios similar to sawn-timber beams	3–6m	20:1 Subject to grade of timber, width and spacing of joists

2.2.3.2
Concrete slab systems

Properties of concrete slab systems

Beam type	Comments	Typical span range	Typical span/depth ratio
Flat slab	**1** Flat slabs contain no downstand beams, providing a continuous flat soffit. They are used in many situations, particularly commercial developments **2** Flat soffits provide easy integration of services, as there is no requirement to divert pipes and ductwork under downstand beams **3** Flat soffits require simple formwork and reinforcement detailing, making them easier and quicker to construct than other forms of concrete-slab construction **4** The self weight of a flat slab can be reduced by inserting void formers within the depth of the slab; these have negligible impact on the structural capacity of the element	6–10m	Multi-span slabs 26:1
Beam and slab (one-way span)	**1** Beam and slab is a traditional form of reinforced-concrete slab and incorporates either one- or two-way spanning slabs **2** The downstand beams reduce the deflection of the system in comparison to a flat-slab solution	5–10m	Multi-span slabs 28:1 Single-span slabs 24:1
Beam and slab (two-way span)	**1** More efficient than one-way spanning slabs for longer spans **2** Requires more formworks to construct downstand beams in both directions	7–12m	Multi-span slabs 35:1 Single-span slabs 32:1
Waffle slab (aka coffered slab; integral)	**1** 'Waffles' create a lightweight reinforced-concrete slab solution **2** Waffle slabs span in two directions, therefore the ratio of the spans in the x and y directions affects the efficiency of the slab. A square bay generally provides the optimum efficiency **3** They can be left exposed as a final finish, thus omitting the need for additional suspended ceilings. This requires the concrete finish to be of a higher than normal standard **4** Waffle-form moulds are typically more expensive than traditional reinforced-concrete formwork, and the reinforcement is more complicated to fix	6–11m	Multi-span slabs 22:1 Single-span slabs 21:1
Ribbed slab (integral)	**1** Lightweight long-span reinforced-concrete slab solution **2** They can be left exposed as a final finish, thus omitting the need for additional suspended ceilings. This requires the concrete finish to be of a higher than normal standard **3** Ribbed form moulds are typically more expensive than traditional reinforced-concrete formwork, and the reinforcement is more complicated to fix	6–11m	Multi-span slabs 22:1 Single-span slabs 20:1

Properties of post-tensioned reinforced concrete slabs

The primary conceptual design aspects of post-tensioned concrete design are listed below:

Post-tensioning is a specialized construction process as described in the post-tensioned reinforced concrete beams section previously.

Post-tensioned concrete slabs can span longer distances than traditional reinforced-concrete slabs.

Slab depths are reduced, leading to less material and therefore less load owing to self weight.

Post-tensioned slabs limit the flexibility of being able to cut holes into a floor system post-construction, owing to the risk of cutting through the post-tensioning tendons.

Beam type	Comments	Typical span range	Typical span/depth ratio
Post-tensioned flat slab	**1** Flat slabs contain no downstand beams, providing a continuous flat soffit. They are used in many situations, particularly commercial developments **2** Flat soffits provide easy integration of services as there is no requirement to divert pipes and ductwork under downstand beams **3** Flat soffits require simple formwork and reinforcement detailing, making them easier and quicker to construct than other forms of concrete-slab construction **4** The self weight of a flat slab can be reduced by inserting void formers within the depth of the slab; these have negligible impact on the structural capacity of the element **5** Additional checks need to be made to ensure vibration limits are achieved	6–11m	Multi-span slabs 22:1 Single-span slabs 30:1
Post-tensioned waffle slab	**1** Waffles create a lightweight reinforced-concrete slab solution **2** Waffle slabs span in two directions, therefore the ratio of the spans in the x and y directions affects the efficiency of the slab. A square bay generally provides the optimum efficiency **3** They can be left exposed as a final finish, thus omitting the need for additional suspended ceilings. This requires the concrete finish to be of a higher than normal standard **4** Waffle-form moulds are typically more expensive than traditional reinforced-concrete formwork, and the reinforcement is more complicated to fix	8–13m	Multi-span slabs 26:1
Post-tensioned beam and slab	**1** Post-tensioned beam-and-slab systems commonly comprise a wide banded beam with a beam width approximately span/5. **2** The downstand beams reduce the deflection of the system in comparison to a flat-slab solution	6–13m	Multi-span beam 22:1 Multi-span slab 40:1 Single-span beam 20:1 Single-span slab 35:1

3
Structural prototypes

3.1
Form finding

During the twentieth century, architects and engineers both developed ways of designing complex structural forms by experimenting with physical models and through borrowing from structures found in nature. Finding and creating new structural forms was accomplished by extracting geometric information from physical models, in particular three-dimensional compressive surfaces – shells – or three-dimensional tensile surfaces – membranes. With the advent of computer-aided design along with an increased knowledge of the behaviour of materials, a variety of approaches to form finding can now be pursued using computer programs to calculate optimum structural solutions for given geometric parameters.

Suspension models

A 'catenary' curve is derived from hanging a chain or a cable that, when supported at each end, is allowed to bend under its own weight. In the case of a suspension bridge, the cables that are stretched between the masts form a catenary curve; however, once the cables become loaded (by hanging a deck from vertical cables placed at regular intervals) the curve becomes almost parabolic. When a catenary curve is inverted, it forms a naturally stable arch. Arches produced in this way are structurally efficient, since the thrust into the ground will always follow the line of the arch.

To generate compressive, shell-like structures, a net or fabric is suspended from a set of points and then fixed in position by saturating it with plaster and/or glue. This is then flipped over (mirrored horizontally) to create a thin shell-like form. Owing to their structural efficiency, these forms may be described as 'minimal surfaces'.

Soap films

Soap bubbles (see section 1.4) are physical illustrations of a minimal surface. A minimal surface is more properly described as a surface with equal pressure on the inside and the outside. A film obtained by dipping a wire frame contoured closed shape into a soapy solution will produce a minimal surface.

Structural application

Designers such as Frei Otto and Heinz Isler have used form-finding structural prototypes as both design and engineering tools. In the case of Otto – and specifically his work with soap films – these models were painstakingly photographed, logged, mapped and drawn, generating profiles for latterly realized projects. Heinz Isler, whose interest was in optimally engineered thin reinforced concrete shells, regularly used physical scale models to generate surface geometries. These reverse-engineered plaster models were very accurately measured on a custom rig, with the subsequently plotted profiles used as the basis for his large-scale 'catenary' shells.

Virtual form finding

Form-finding software is now widely available as a design and analysis tool and is no longer solely the domain of the professional engineering office. Form-finding software is based on principles such as the geometric optimization of the soap-film modelling techniques pioneered by Frei Otto. Typical form-finding software contains a range of procedural geometric transformations as well as ascribable properties for the constituent material construction and arrangement, which may include fabric type, steel cabling and connectors. The virtual model can then be subject to prestress and live load simulations. While there is no question of the value of these excellent new tools, which allow for fast iterative modelling, there are still good arguments for physical prototyping. The physical scale model as an analogue of the final physical construction has much to tell the designer, not least in relation to material behaviour and project-specific constructional and assembly issues.

1

2

1 Hanging nets
Antoni Gaudí's models explored the design of vaulted compression structures using the same principle as the catenary curve, by hanging weights from flexible nets and then inverting the resultant forms

2 Suspension model
Structural model made to establish the form of the arches for a new train station in Stuttgart, Germany, 2000, by Christoph Ingenhoven and Partner, Frei Otto, Büro Happold, and Leonhardt and Andrae

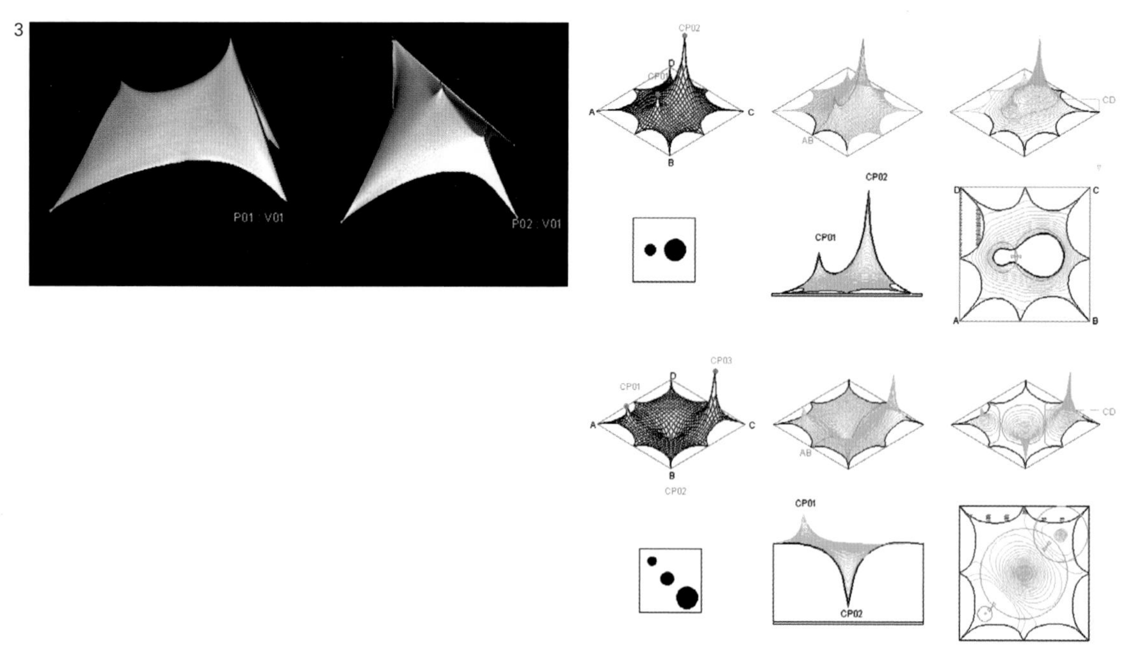

3 Control points
Images created using form-finding software for the design of membrane structures. Control points (CP) are used to create space. The program operates in such a way that when a force is applied to one point the load of the force is distributed homogeneously so that the membrane is always under tension to produce a smooth transition between points.

4 Soap-film model
Model by Frei Otto for the design of a membrane structure using soap film on a wire-bounded framework. This is both a minimal and an anticlastic surface, which can be graphically described as a 'double-ruled' surface, i.e. one that can be described using a grid of straight lines.

5

5 Ice shells
Heinz Isler designed a technique whereby fabric was draped over masts and then saturated with water. In freezing temperatures the membranes solidified and the masts could be removed, forming 'ice shells'. Shown here is an image of ice shells constructed at Cornell University, Ithaca, NY, in 1999 by Dr. Mark Valenzuela and Dr. Sanjay Arwade, with the assistance of undergraduates from Dr. Valenzuela's Modern Structures class.

6

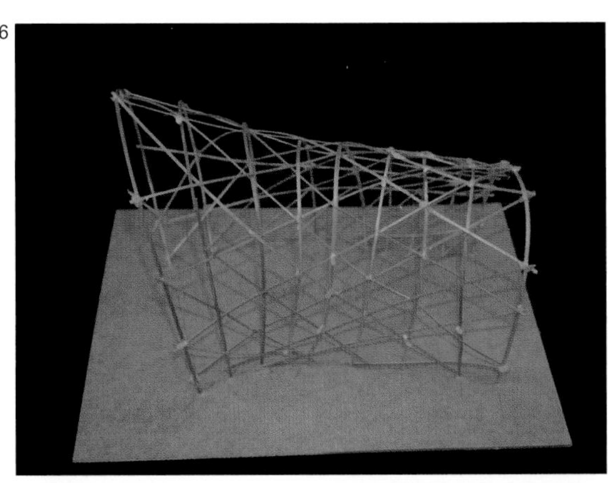

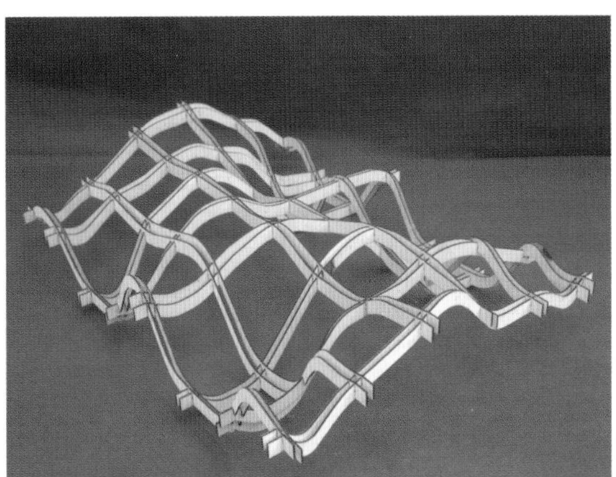

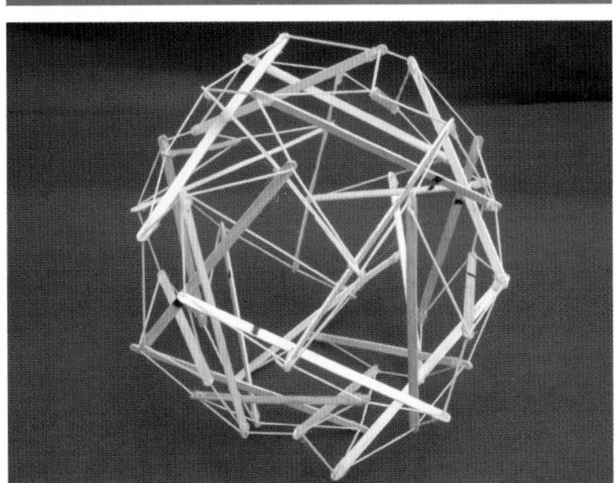

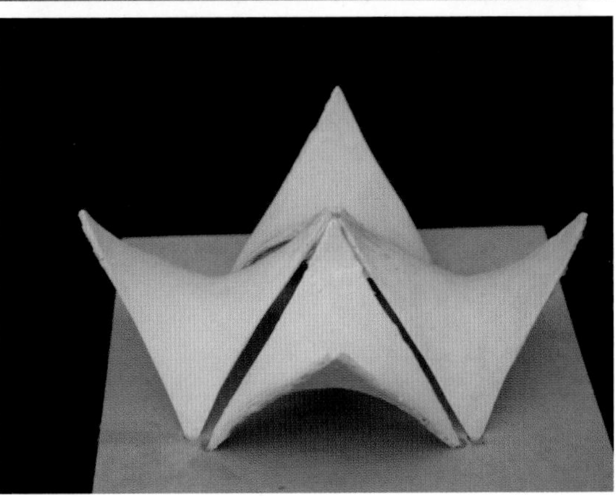

6 Modelling techniques
Structural models can employ a range of form-finding techniques according to the properties of the materials used, as shown in the examples here. All models by second-year undergraduate students at the School of Architecture, University of Westminster, London, 2007–9.

Left to right, top to bottom:

Gridshell vault, formed using (elastic) timber strips that are held in tension and fixed at the base of the model.

Complex surface built up of laser-cut profiles in an interlocking grid

Interlocking cardboard profiles used to model a formwork core

Disposable sticks and elastic bands employed to model a collapsible tensegrity dome

Paper ribbons folded and interlocked to generate a regular solid

Hyperbolic parabolas (saddle shapes) generated by saturating a hung fabric with plaster

3.2
Load testing

Load testing has always been a critical part of structural design development. While the prediction of the behaviour of materials and construction elements may be calculated mathematically and with computer models (such as Finite Element Analysis and Computational Fluid Dynamics, see pages 106–9), much can be learned by prototyping and observation. The first time it was understood that reinforced concrete could flex and bend under load was on the completion of Berthold Lubetkin's Penguin Pool at London Zoo in the 1930s.

As can be seen from Robert Stevenson's work on the Bell Rock Lighthouse, there is evidence that the use of prototype models was paramount to the resolution of successful structural design to resist the enormous power of the sea. Similarly, monolithic, compressive vaults and domes have, from Gothic times, required innovative construction techniques and materials that are still under constant development. This section is also illustrated by a set of practical, problem-solving exercises, showing examples of a considerable variety of resolutions.

1

1
The spiralling, reinforced-concrete ramp of Lubetkin's Penguin Pool at London Zoo, under construction in 1933

2 Bell Rock Lighthouse
The design of the Bell Rock was the culmination of knowledge gained from the construction of previous lighthouses (many of which had failed) and from prototyping with scale models. John Smeaton had built the Eddystone Lighthouse in 1759, pioneering the use of stone. Not only were the stones 'dovetailed' to interlock with each other, but they also employed wedges similar to the dowels in a 'scarf' joint. The ideal profile to resist the enormous impact from wind and waves was found to be parabolic in shape; Robert Stevenson and John Rennie are known to have built scale models against which they would throw buckets of water.

Left:
Photograph of Bell Rock Lighthouse showing the parabolic curve at the base

Below left:
Section through the interlocking stone blocks at foundation level

Below:
Models of the construction details. Held at the Museum of Scotland, Edinburgh

2

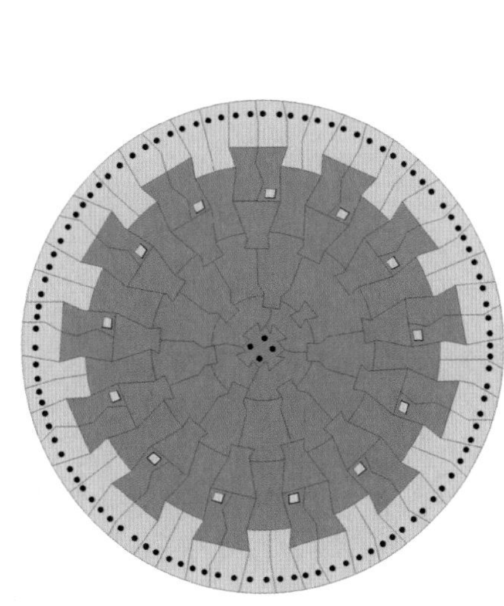

3

3 Thin-shell monolithic domes
Modern (lightweight) materials technology linked to the use of air-supported formworks has greatly improved the efficiency and practicality of casting concrete domes (which are similar in shape and structure to an eggshell). Inspired by prototypes developed by Félix Candela, Pier Luigi Nervi and Anton Tedesko among others, shown here is a project by Dr. Arnold Wilson at the Brigham Young University Laboratories, Idaho, USA, to load test a thin-shell concrete dome. Using air-supported form technology (made from nylon-reinforced vinyl, which is left in place as a watertight finish), the dome is formed using polyurethane foam and sprayed (reinforced) concrete.

Above:
Inflatable formworks, showing reinforcing

Left:
Load testing a dome

4

4 Brick vaults
Inspired by the work of Eladio Dieste and others, the Vault201 prototype vault was built by MIT architecture students at the Cooper-Hewitt National Design Museum, New York. The vault spans 4.88m, is 4cm thick, and uses 720 bricks. The curvature of the vault is composed of splines that vary in profile but are fixed in length in order to keep an equal coursing pattern and to save custom-cutting too many bricks. In the end, as a result of prototyping, a taxonomic system of three different brick modules was developed.

To quote the students: '1) learn from building, 2) analyse and abstract as rules, and 3) re-embed into the design process.'

(See http://vaulting.wordpress.com/ for a full account of this project.)

3.2
Load testing

The following illustrations are taken from first-year undergraduate student projects conducted at the University of Westminster, London, UK, from 2009 to 2011. Students were introduced to common construction materials, fabrication processes and workshop practices and were then asked to design and build a 1:1 scale object in order to solve a specific structural problem. Prototyping took the form of sketching, modelmaking and experimenting with materials, and students learned how the act of 'making' can form an integral part of the design process. Objects were assessed according to structural efficiency (lightness), craftsmanship, construction details, and the innovative use of materials.

5

a

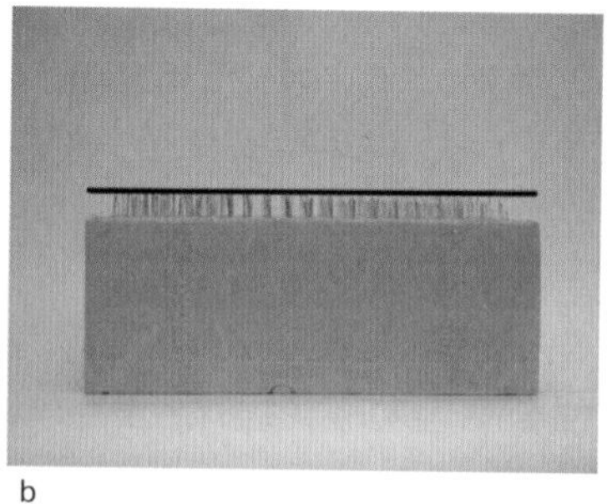

b

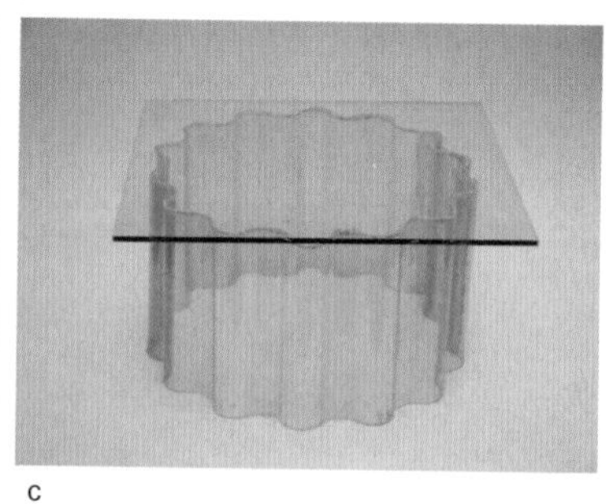

c

d

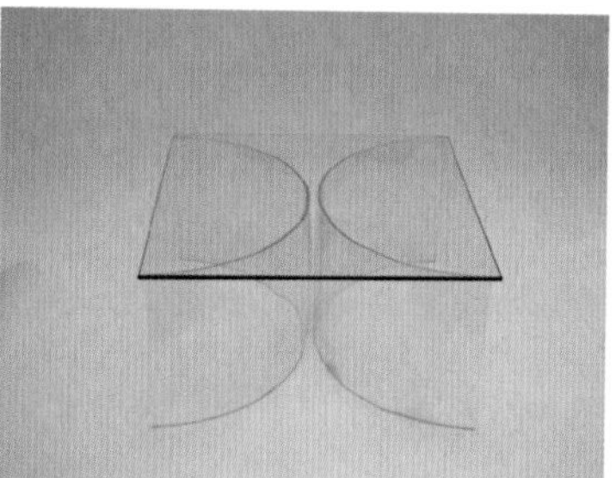

e

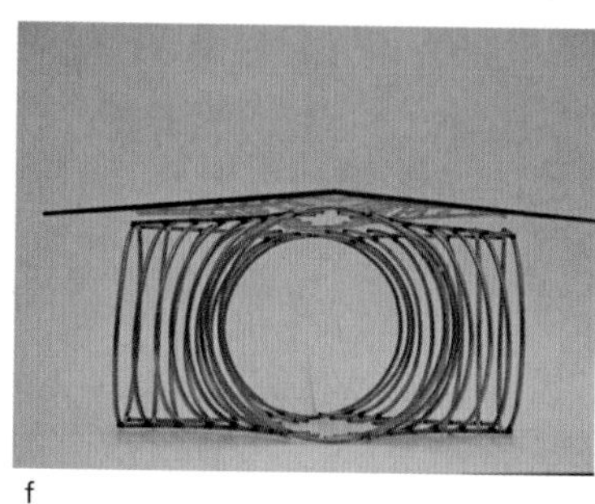

f

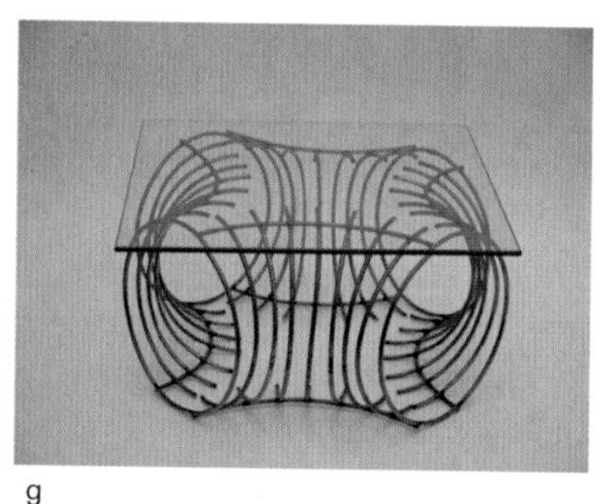

g

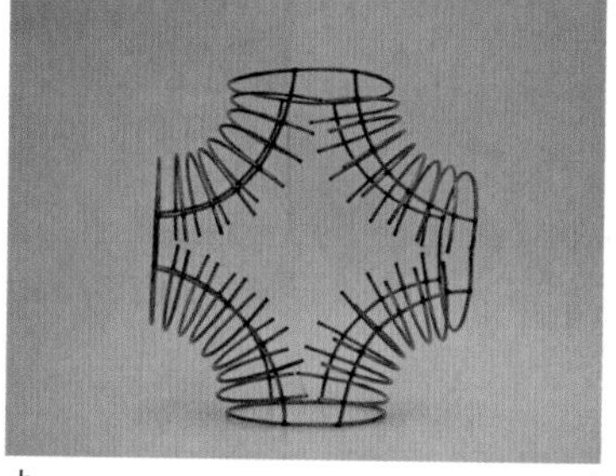

h

i

j

k

l

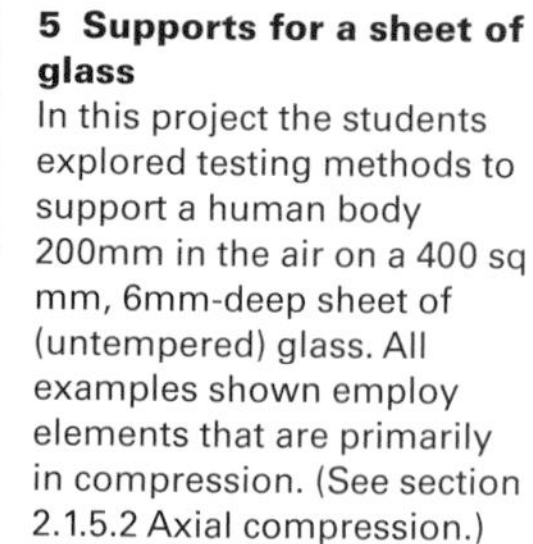

5 Supports for a sheet of glass
In this project the students explored testing methods to support a human body 200mm in the air on a 400 sq mm, 6mm-deep sheet of (untempered) glass. All examples shown employ elements that are primarily in compression. (See section 2.1.5.2 Axial compression.)

a, b
Multiple, point-loaded structure exploring iteration and scale

c–e
Stiffness achieved using corrugation and stability through the use of a circular plan

f–h
The arch and the cantilever principle combined

i, j
Pointed arches used as colonnades

k, l
A pointed arch, perforated for lightness, acting as a portal frame

m
Students testing their support structures

m

6

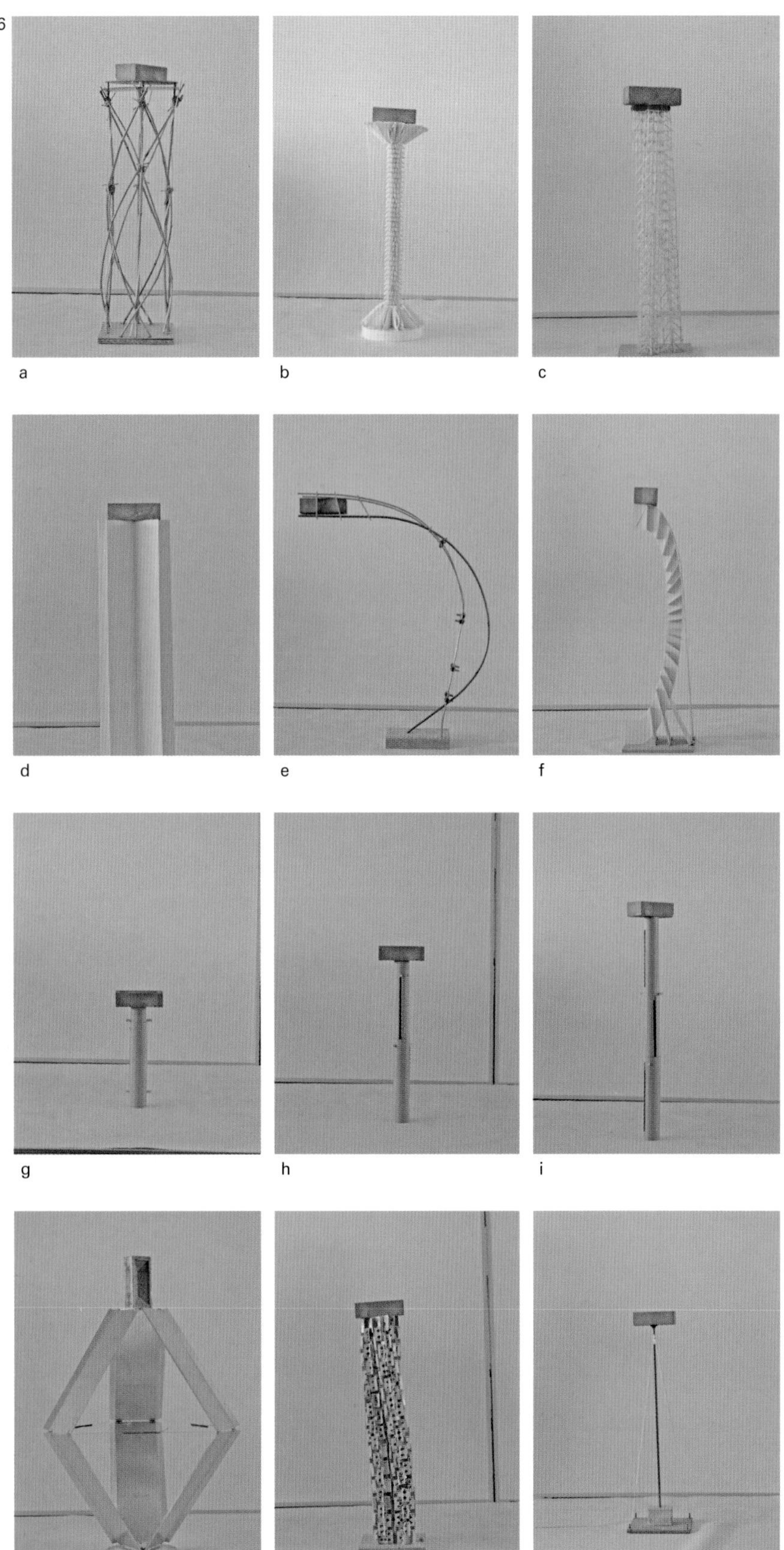

a b c d e f g h i j k l

6 Brick-supporting plinth
The brief was to design a column that could support a single Fletton brick at a height of one metre, without bending, buckling or rotating. The load was considered primarily to be vertical, though the plinth should resist torsion (see section 2.1.4.1 Stress: Element under torsion). Maximum footprint was set at 250 x 250mm. These efficient minimal structures were to be designed to fail under the load of two bricks.

a
This project set out to explore the structural potential of the double helix by employing elements made from a stiff material with the capacity for elastic deformation – in this case, bamboo. Torsive forces were applied in order to twist opposing elements in opposite directions; they were then locked at either end so that the forces cancelled each other out. This produced an extremely rigid structure with a high strength-to-weight ratio

b
This project consisted of a mast that was made up of multiple, folded (paper) elements slotted around a cylindrical core. Rigidity was achieved through a system of bracing that would resist torsional movement by tensioning lever arms at the top and base, using a network of triangulating wires

c
A lightweight, compressive lattice consisting of three masts that were intertwined for stability

d
A monolithic, planar structure whose form was derived by extruding from a simple plan. A series of ribs was connected (critically) at the point of rotation. To prevent the thin, planar ribs from buckling under load, they were individually laminated (using foamboard)

e
This project explored the possibility of cantilevering the brick, while at the same time employing a minimum number of primary elements. By using two rods with the capacity for elastic deformation they could be 'laminated' together to act simultaneously in tension and compression to form a rigid structure

f
A single, tapering lever arm was stiffened using a series of ribs, which also acted to stabilize the structure at ground level. The vertical cantilever was completed by tensioning the lever arm back to the base of the structure

g–i
This deployable solution used a telescoping mechanism. A set of cardboard cylinders was slotted so that they could be pegged at various heights

j
This project set out to leave a clear space below the brick while also being deployable. The solution involved using three armatures that were each centrally hinged. The desired height was achieved by tensioning each of the arms to its neighbour with the appropriate length of cable

k
The core of the mast consisted of cards that were stacked and slotted together vertically. Rigidity was achieved by tensioning the top to the base

l
This simple column was stabilized by tensioning cables to the base plate

m
Supported by 200 helium balloons, the brick was held by a perforated, polystyrene beam in order to stabilize and spread the load

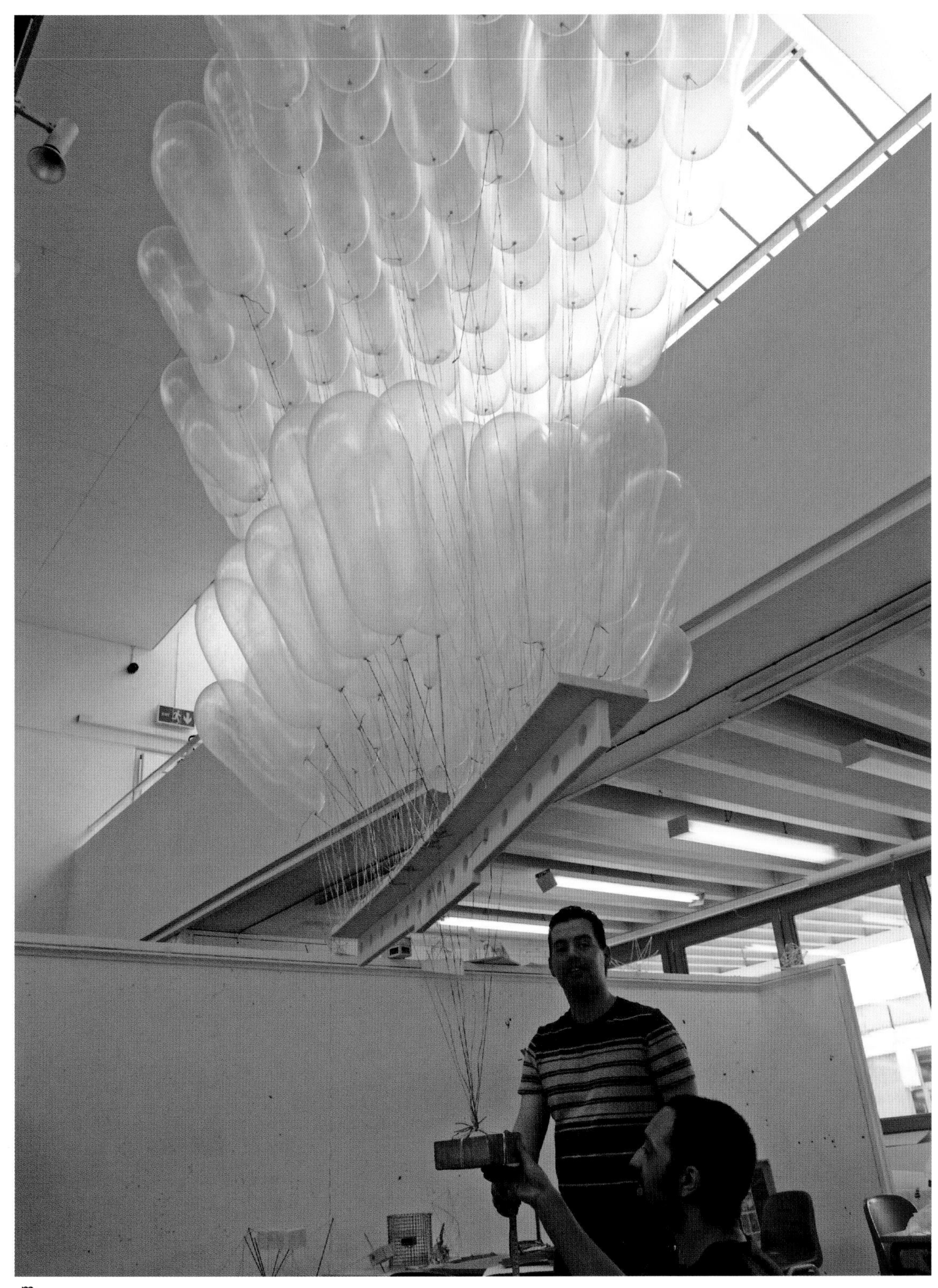

m

7

400mm

400mm

Load from
apple must be
transferred to
ground within
this space

400mm

400mm

a

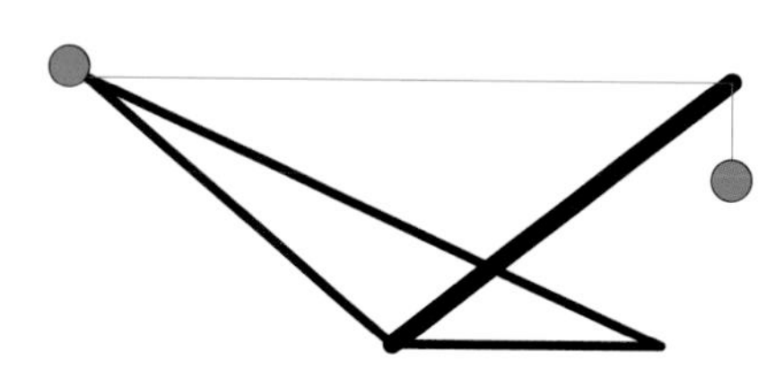

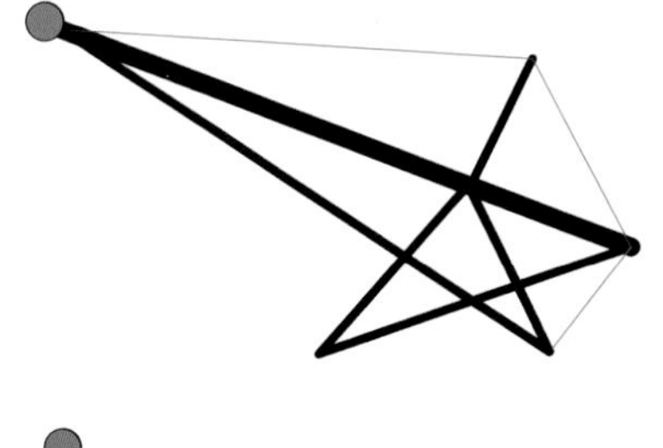

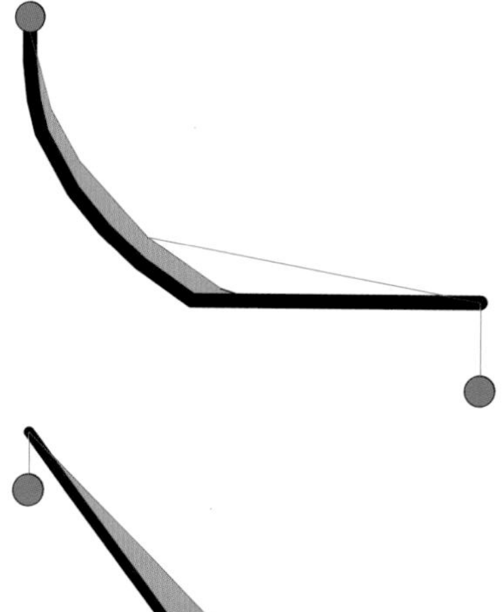

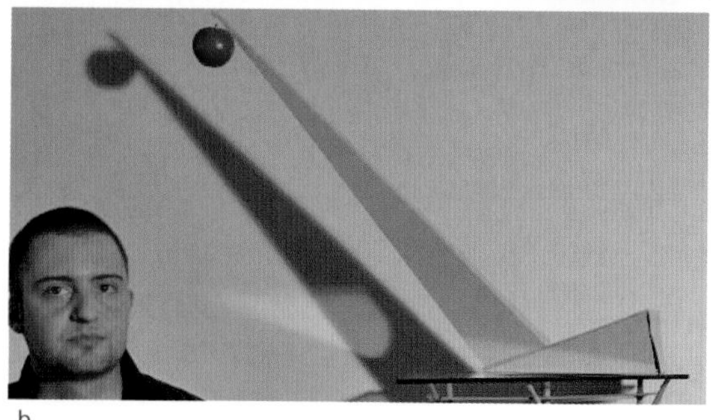

b

7 Cantilever support for an apple
The following illustrations show the results of an exercise to explore solutions to cantilevering an apple 400mm horizontally and 400mm vertically from a 400 x 400mm footprint. The load was considered primarily to be vertical, though the apple should remain stable in the horizontal plane. The diagrams describe the tensile (red) and compressive (black) elements at work in the structures.

a
Diagram explaining the general requirements for each structure

b
Photographs of four selected structures with diagrams describing the tensile (red) and compressive (black) elements

c
The students' solutions to this structural problem were varied and inventive

c

8

8 Glass 'sandwich' panel spanning element
This structural prototype developed by David Charlton at the University of Westminster, London, UK, creates a 'sandwich' panel using traditional incandescent light bulbs close-packed in hexagonal plan formation and bonded to thin sheets of glass using structural silicone. The glass honeycomb-like core created from recycled light bulbs utilizes the relative longitudinal compressive strength of the bulb similar to that of an eggshell (see Section 1.3 Eggshell). The close-packing of the bulbs resists the tendency of the bulbs to buckle (and fracture), providing lateral stability. This novel prototype reminds us of the usefulness of putting distance between the top and bottom chord of a beam, truss or spaceframe, thus creating structural 'depth' with which to 'span'. This prototype also shows how, with thoughtful geometric configuration, compressive strength can be maintained with lightweight and even fragile materials maintaining impressive strength and reducing dead (static) loads.

9

9 Cable net structure
A cable net structure for a DIY version of London's O_2 Arena (formerly the Millennium Dome) was constructed by first-year undergraduate students at the University of Greenwich, London, UK. This 1/36 scale model utilized all the structural attributes of the original, albeit simplified by using eight rather than twelve uprights (compression members) for this mast-supported cable net.

3.3
Visualizing forces

A key development in engineering analysis has been the ability to visualize forces within a 'structural model'. In a process developed at the beginning of the twentieth century, photoelastic modelling allowed scale models fabricated from transparent cast resin to have the internal structural forces made visible. Using two polarizing lenses set each side of a scale model, light is passed through the rig, and birefringence (double refraction) occurs in direct relation to localized stress patterns. Whereas physical models may have been previously used to verify structural calculations, these photoelastic structural 'analogues' allow the designer to simultaneously test and observe structural forces and structures in action.

With the development of Finite Element Analysis (FEA) and application of the Finite Element Method (FEM), graphical computing allows the designer to model a two- or three-dimensional structural system or connection and study the fourth dimensional effects of gravity, static and live loads and other applied structural forces. The advent of inexpensive computing allows a fully integrated Building Information Model (BIM) to be recast or reconfigured with information feedback from FEA analysis and additional dynamic environmental factors such as wind loads, modelled with Computational Fluid Dynamic (CFD) software.

Photoelastic modelling

Photoelastic modelling is an experimental method to determine stress distribution in a material, and is often used for determining stress-concentration factors in complex geometrical shapes. The method is based on the property of birefringence, which is exhibited by certain transparent materials. A ray of light passing through a birefringent material experiences two refractive indices. Photoelastic materials exhibit this property only on the application of stress, and the magnitude of the refractive indices at each point in the material is directly related to the state of stress at that point. A model made out of such materials produces an optical pattern representing its internal stress patterns.

Professor Robert Mark of Princeton University brilliantly illustrates both the method and analytical usefulness of the photoelastic technique in his book *Experiments in Gothic Structure* (MIT Press, Cambridge, MA, 1982), where a series of comparative (sectional) models of some of the great Gothic cathedrals of Europe are photoelastically modelled and subjected to notional live (wind) loads. These live and responsive illustrations of stress patterns in a given structure provide valuable indicative evidence of localized 'hot spots' for study or amelioration. The correlating numerical and algebraic structural calculations, however, must be separately computed.

1

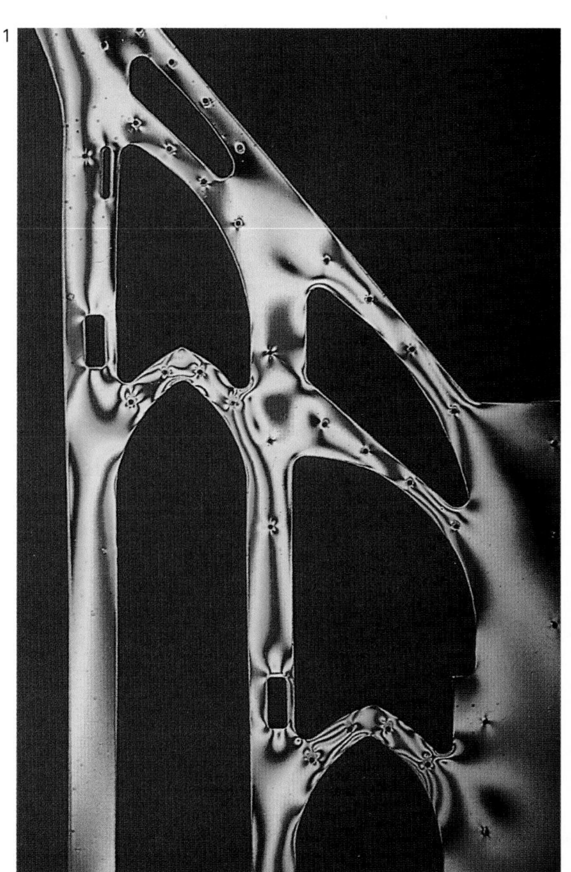

2

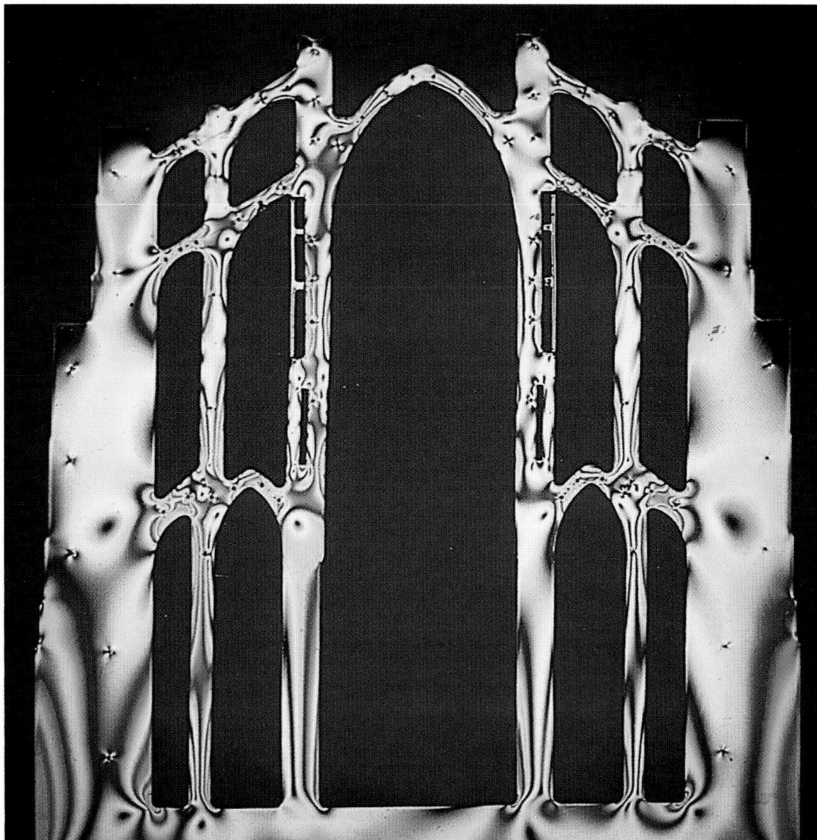

3

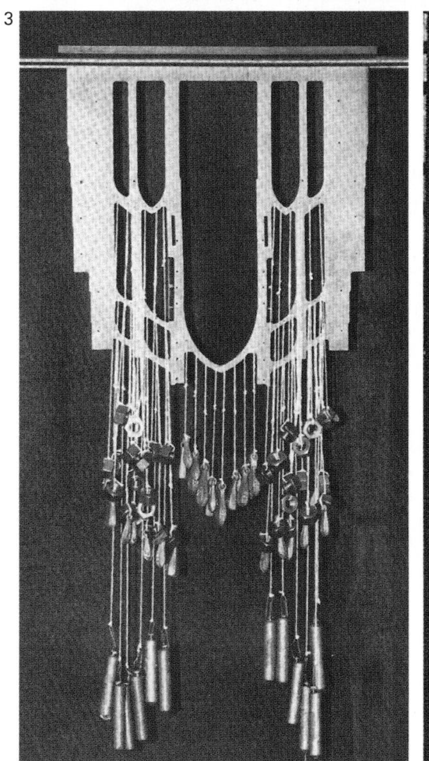

4

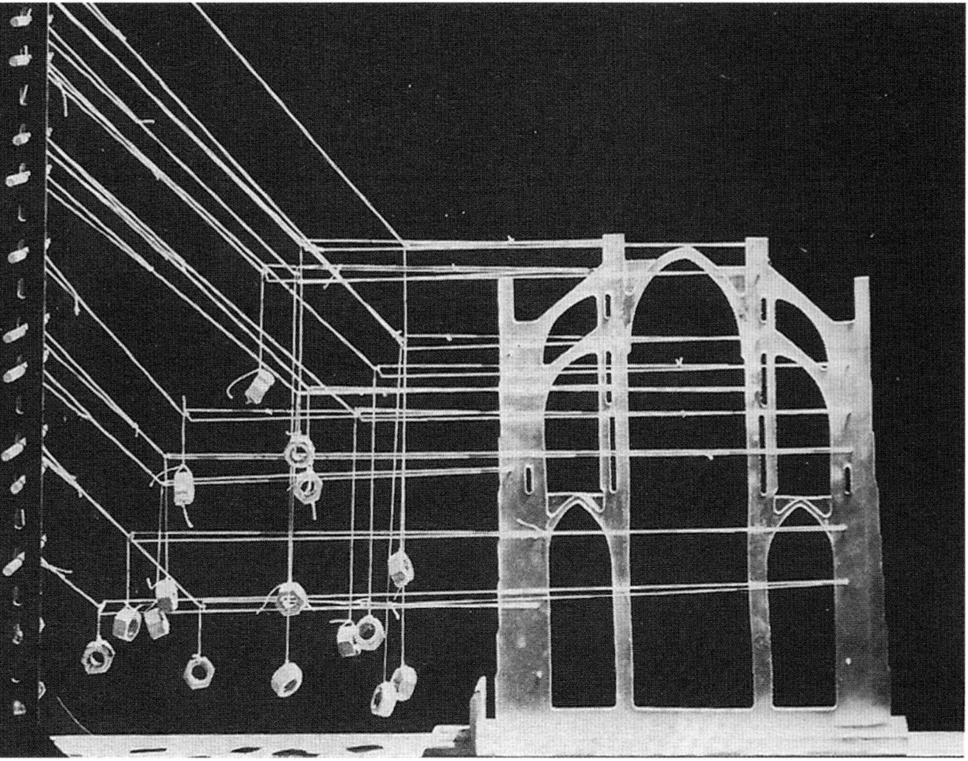

1
Photoelastic model of Bourges Cathedral choir. The photoelastic interference patterns are produced by simulated dead weight (static loading).

2
Photoelastic model of Beauvais Cathedral choir. The photoelastic interference patterns are produced by simulated wind loading.

3
Photograph showing how Professor Mark simulated dead weight (static loading) on a model of Beauvais Cathedral using hanging weights of differing masses, attached to corresponding cross-sectional locations.

4
A live loading model of Amiens Cathedral subjected to simulated lateral wind loading. Vertical wires are attached to the model and evenly weighted.

Finite Element Analysis (FEA)

The first step in using Finite Element Analysis (FEA) is constructing a finite element model of the structure to be analysed. Two- or three-dimensional CAD models are imported into FEA software and a 'meshing' procedure is used to define and break the model up into a geometric arrangement of small elements and nodes. Nodes represent points at which features such as displacements are calculated. Elements are bounded by sets of nodes and define the localized mass and stiffness properties of the model. Elements are also defined by mesh numbers, which allow reference to be made to corresponding deflections or stresses at specific model locations. Knowing the properties of the materials used, the software then conducts a series of computational procedures to determine effects such as deformations, strains and stresses, which are caused by applied structural loads. The results can then be studied using visualization tools within the FEA environment to view and to identify the implications of the analysis. Numerical and graphical tools allow the precise location of data such as stresses and deflections to be identified.

1

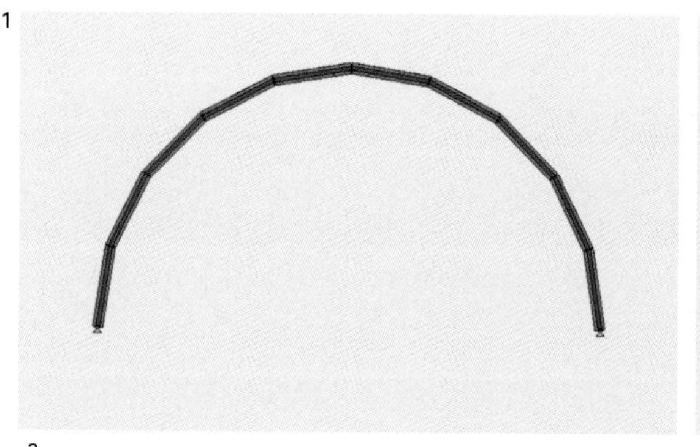

a

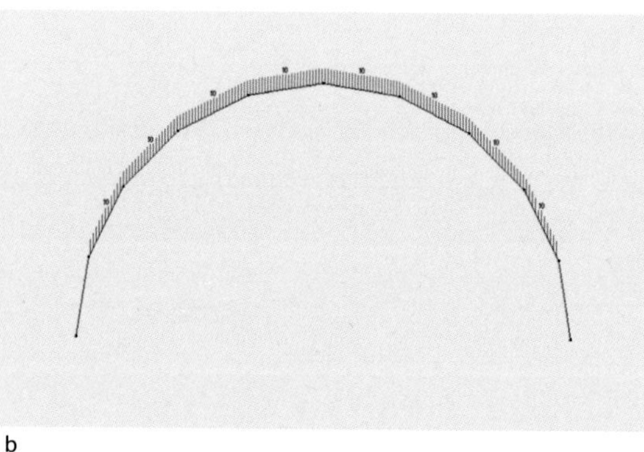

b

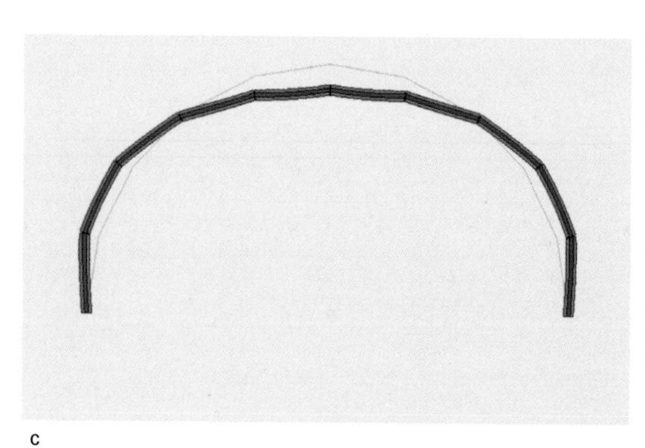

c

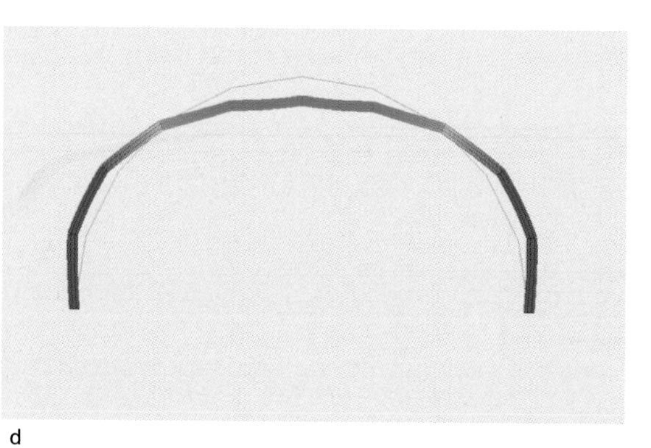

d

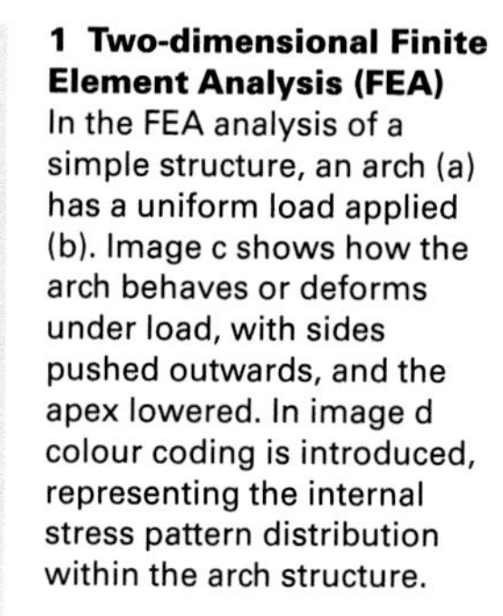

1 Two-dimensional Finite Element Analysis (FEA)
In the FEA analysis of a simple structure, an arch (a) has a uniform load applied (b). Image c shows how the arch behaves or deforms under load, with sides pushed outwards, and the apex lowered. In image d colour coding is introduced, representing the internal stress pattern distribution within the arch structure.

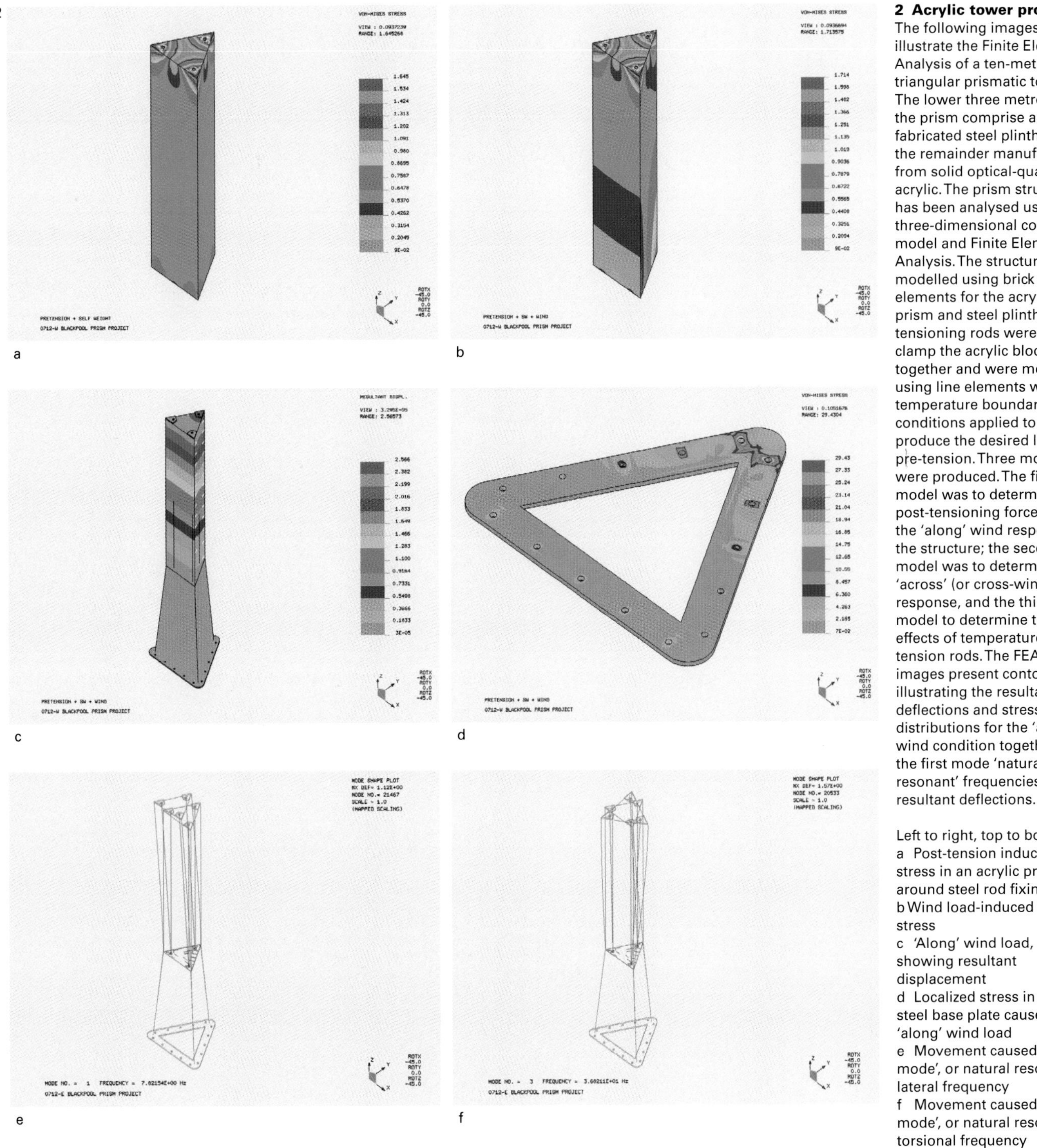

2 Acrylic tower project
The following images illustrate the Finite Element Analysis of a ten-metre-tall triangular prismatic tower. The lower three metres of the prism comprise a fabricated steel plinth with the remainder manufactured from solid optical-quality acrylic. The prism structure has been analysed using a three-dimensional computer model and Finite Element Analysis. The structure was modelled using brick elements for the acrylic prism and steel plinth. Steel tensioning rods were used to clamp the acrylic blocks together and were modelled using line elements with temperature boundary conditions applied to produce the desired level of pre-tension. Three models were produced. The first model was to determine the post-tensioning forces and the 'along' wind response of the structure; the second model was to determine the 'across' (or cross-wind) wind response, and the third model to determine the effects of temperature on the tension rods. The FEA images present contour plots illustrating the resultant deflections and stress distributions for the 'along' wind condition together with the first mode 'natural resonant' frequencies and resultant deflections.

Left to right, top to bottom:
a Post-tension induced stress in an acrylic prism around steel rod fixings
b Wind load-induced acrylic stress
c 'Along' wind load, showing resultant displacement
d Localized stress in the steel base plate caused by 'along' wind load
e Movement caused by 'first mode', or natural resonant lateral frequency
f Movement caused by 'first mode', or natural resonant torsional frequency

Computational Fluid Dynamics (CFD)

The Navier-Stokes equations, named after Claude-Louis Navier and George Gabriel Stokes, are a set of equations that describe the motion of fluid substances such as liquids and gases. The equations are a dynamical statement of the balance of forces acting at any given region of the fluid. The various numerical approaches to solving the Navier-Stokes equations are collectively called Computational Fluid Dynamics, or CFD. When translated into a graphical format, the motion of the fluids can be seen as particles moving through space. CFD can then be used to simulate wind dynamics – speed and direction – in and around buildings. The architect is able to explore variations in design that can, for example, improve natural ventilation or minimize excessive downdrafts from tall buildings. Using in-built or referenced weather data, this analytical computer software allows the user to model and overlay annual wind speed, frequency and direction, directly on top of a design model, helping the designer develop strategies for natural ventilation, wind shelter and appropriate structural resistance.

1

1 CFD flow vector analysis section
CFD flow vector analysis showing air movement and velocity in a cross-sectional view of an urban block.

2

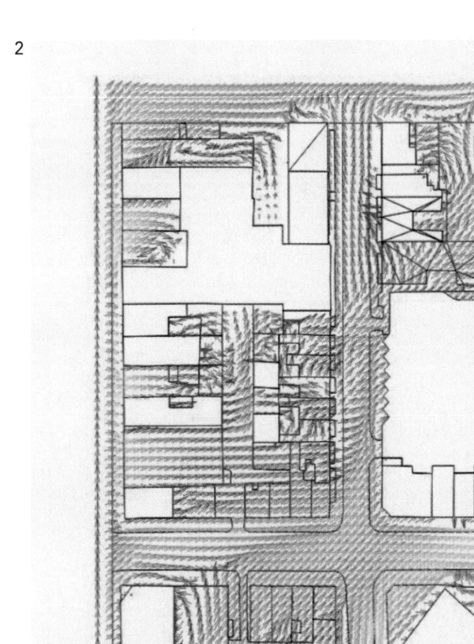

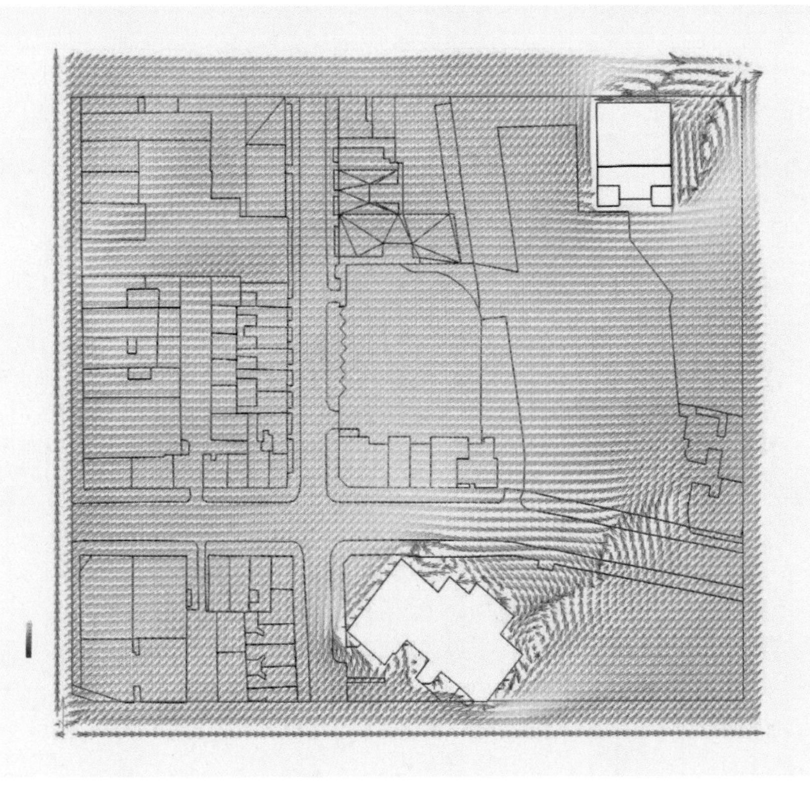

2 CFD flow vector analysis plan
CFD flow vector analysis showing air movement and velocity at two heights above an urban block. Note the prevailing southwesterly wind flow and the turbulence and vortex shedding around the tall building at the centre bottom of the images.

4
Case studies

4.1 Introduction

In attempting to describe and explain structures and structural principles the case studies that seem the most useful are often highly individualized 'one offs', exemplary artefacts of unique individuals whose work was formed as a part of a larger philosophical approach to societal needs, both contemporary and for the future. We see this approach, albeit in vastly different ways and means, in the work of structural innovators such as Pier Luigi Nervi, Richard Buckminster Fuller and Konrad Wachsmann, to name but three. All of these structural artists produced compelling and prototypical projects experimenting with new construction processes, fabrication methods, new geometric configurations and programmatic determinants. The work of these structural pioneers also tested (or completely circumvented) the limits of contemporary engineering and architectural orthodoxy, presenting new models for the production of our built environment, the principles of which we still struggle to fully understand, let alone wholeheartedly embrace. While these individuals are now figures from the last century, their experimental work can be usefully understood as presenting a beautiful diversity of approach for new ways of making the world. It is in this context that the exhibition mounted at the Architecture Museum, Munich, in 2010, 'Wendepunkte im Bauen – Von der seriellen zur digitalen Architektur', revisited Konrad Wachsmann's seminal book of 1961 *The Turning Point of Building* and staged a show of how the legacies of work by figures like Wachsmann can be re-read, thoughtfully assimilated and, with the addition of new digital fabrication tools, provide the fuel for some new kinds of architecture and engineering that usefully (and delightfully) serve society. If it is unclear whether the likes of Nervi, Fuller, Wachsmann (and even Jean Prouvé and Frei Otto) are engineers,

architects, builders or artists then that surely is the point. The relationship between the perception of the architect and the engineer can and has been a problematic one, with mutual misunderstanding from the less talented (or officious) of both arts. Unhelpful internecine squabbles about which professional body lays claim to which talent is irrelevant, except to say that these professions were slow to claim any of the aforementioned individuals as one of their own, their prodigious but aberrant talents dismissed or, worse, treated with benign neglect by their 'professions'.

The case studies are simply intended as a collection of structural diagrams, self-illustrating structural and material investigations realized as architecture. These range from the highly unusual concrete truss roof at Chiasso by Robert Maillart from 1924 (which can be seen as a direct translation or materialization of the structure's very own structural analysis) to Antón García-Abril's whimsical structural experiment the Hemeroscopium House of 2008.

The case studies are shown within the context of the general work and impact of their creators, and are presented chronologically. With just over half of the examples derived from the second half of the twentieth century, this section also includes significant structures from the latter half of the nineteenth century and state-of-the-art projects from the beginning of the twenty-first.

4.2
1850–1949

4.2.1
Viollet-le-Duc's innovative engineering approaches

Structural description
Rib vaulting

Engineer
Eugène Viollet-le-Duc
(1814–1879)

1

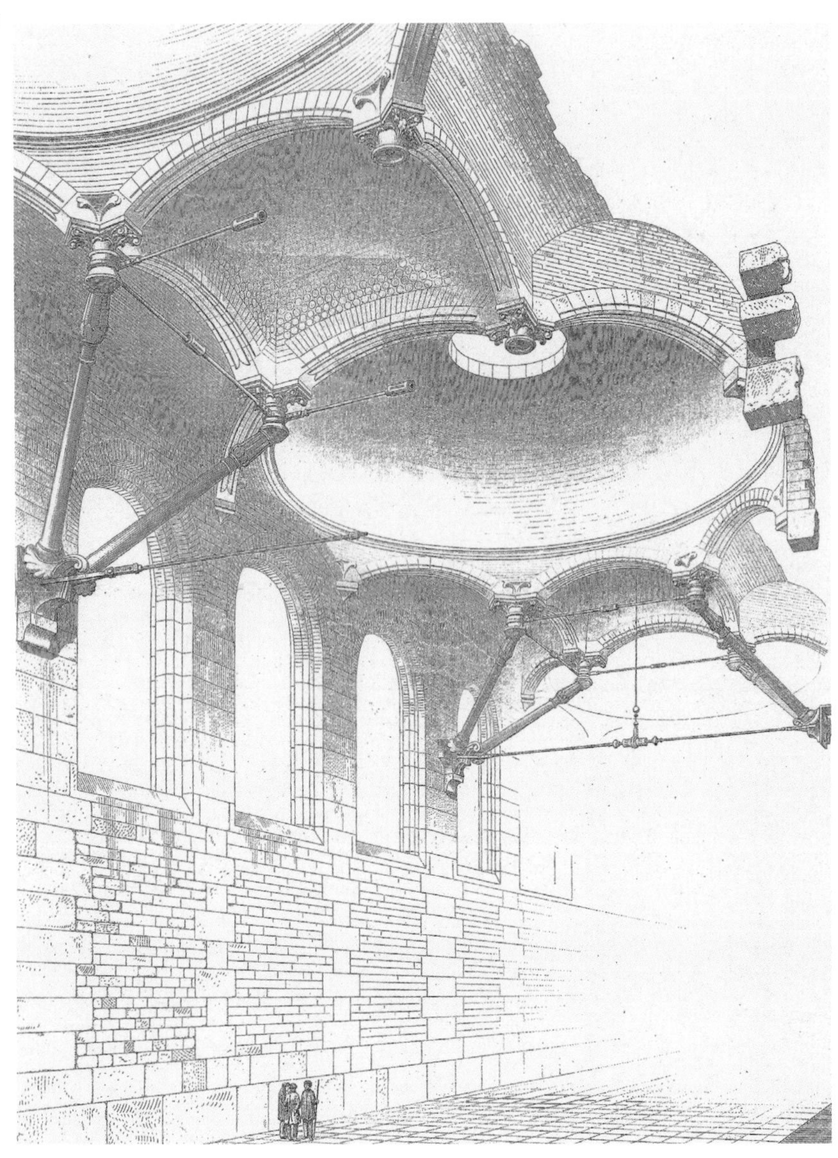

Viollet-le-Duc was responsible for a series of major restorations to medieval buildings, and produced two significant illustrated dictionaries of architecture. He was considered an artist, a scientist, an engineer, an archaeologist and a scholar.

Viollet restored Notre Dame Cathedral, Hôtel de Cluny and other well-known medieval buildings in Paris as well as the cathedrals of Amiens, Saint-Denis and Lausanne (for which he was awarded a medal by an international jury) and numerous city halls and chateaux. He considered that the restoration of Gothic architecture required a deep understanding of, and respect for, the structural engineering from which much of its beauty was derived, but was not afraid to reinterpret a brief. He wrote that restoration is a 'means to reestablish (a building) to a finished state, which may in fact never have actually existed at any given time'.[1]

In several unrealized projects for new buildings, Viollet determined that it was appropriate to apply the construction and materials technology of the day (such as cast iron) to the structural logic and forms of the Gothic period. He also explored natural forms, such as leaves and animal skeletons, and used the wings of bats as an influence for the design of vaulted roofs.

1 Viollet-le-Duc, E., *The Foundations of Architecture: Selections from the Dictionnaire raisonné*, New York: George Braziller, 1990, p. 170

2

1, 2
Compositions in masonry and iron. From E. Viollet-le-Duc, *Entretiens sur l'Architecture*, Paris, 1863

4.2.2
St. Pancras Railway Station Shed

Structural description
Wrought-iron barrel vault roof with cast-iron columns

Location
London, England

Completion date
1869

Plan dimensions
210m long x
73m arch span wide

Height
30.5m

Engineers
William Henry Barlow (1812–1902) and Rowland Mason Ordish

Contractor
The Butterley Company

1

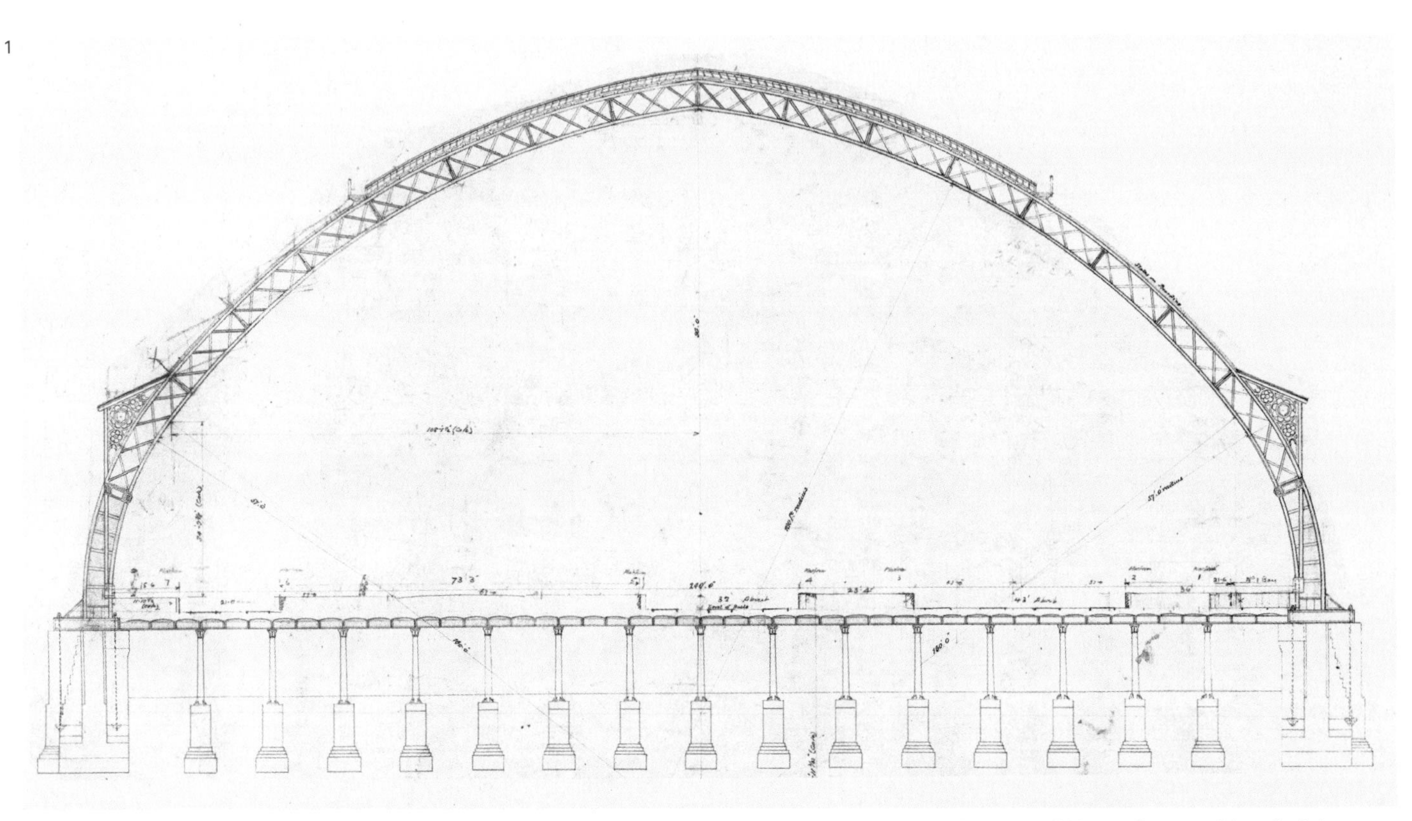

1
St. Pancras Station, the meeting of the styles: section

The initial plan of the station was laid out by William Henry Barlow. Barlow modified the original plans by raising the station 6 metres on 720 iron columns, thus providing a usable undercroft space and also allowing the approach tracks to cross the Regent's Canal on a bridge rather than a tunnel. A space for a hotel fronting the shed was included in the plan, and the competition for its design was won by George Gilbert Scott with a brick Gothic Revival building.

With a covered area of 17,000 square metres, William Barlow's train shed is still considered to be one of the largest enclosed spaces in the world.

2

3

2
Station under construction

3
St. Pancras Station, interior view

4.2.3
Eiffel Tower

Structural description
Steel pylon or lattice tower

Location
Paris, France

Completion date
1889

Height
324m

Plan dimensions
125m x 125m

Engineers
Gustave Eiffel (1832–1922), Maurice Koechlin and Emile Nouguier

Contractor
Gustave Eiffel (Eiffel & Cie)

Architect
Stephen Sauvestre

For the Universal Exhibition of 1889, a date that marked the centenary of the French Revolution, the French *Journal Officiel* launched a major competition to 'study the possibility of erecting an iron tower on the Champ-de-Mars. The tower would have a square base, 125 metres on each side and 300 metres high'. The proposal by entrepreneur Gustave Eiffel, engineers Maurice Koechlin and Emile Nouguier and architect Stephen Sauvestre was chosen. In 1884, Gustave Eiffel had registered a patent 'for a new configuration allowing the construction of metal supports and pylons capable of exceeding a height of 300 metres'.[1] The company was aiming to achieve the iconic height of 1,000 feet (more precisely, 304.8 metres). For the competition, Stephen Sauvestre was employed to transform what was essentially a large pylon into a decorative, functional structure. He proposed stone pedestals to dress the legs, monumental arches to link the columns and the first level, large glass-walled halls on each level and a bulb-shaped design for the top.

The curvature of the uprights was designed to offer the most efficient wind resistance possible, as Eiffel explained: 'Now to what phenomenon did I have to give primary concern in designing the Tower? It was wind resistance. Well then! I hold that the curvature of the monument's four outer edges, which is as mathematical calculation dictated it should be ... will give a great impression of strength and beauty, for it will reveal to the eyes of the observer the boldness of the design as a whole. Likewise the many empty spaces built into the very elements of construction will clearly display the constant concern not to submit any unnecessary surfaces to the violent action of hurricanes, which could threaten the stability of the edifice.'[2]

The greatest difficulty in erecting the tower was the connection of the four main pillars at the first-floor level. The pillars sprang at a precise angle from bases that were 80 metres apart to connect with the first floor at a height of 50 metres.

All of the construction elements were fabricated in Eiffel's factory located on the outskirts of Paris. Each of the 18,038 sections used to construct the tower was traced out to an accuracy of a tenth of a millimetre, and they were then put together using temporary bolts to form prefabricated sections of around 5 metres in length.

On site, the bolts were replaced one by one with a total of 2,500,000 thermally assembled rivets, which contracted during cooling to ensure a very tight fit.

The pillars rest on concrete foundations installed a few metres below ground level on top of a layer of compacted gravel. Each corner edge rests on its own supporting block, applying to it a pressure of 3 to 4 kilograms per square centimetre, and each block is joined to the others by underground walls.

In all, the construction weighs 10,100 tonnes. Between 150 and 300 workers were on site at any one time.

1, 2 Eiffel, G., Excerpt from an interview in the French newspaper *Le Temps*, February 14, 1887

1

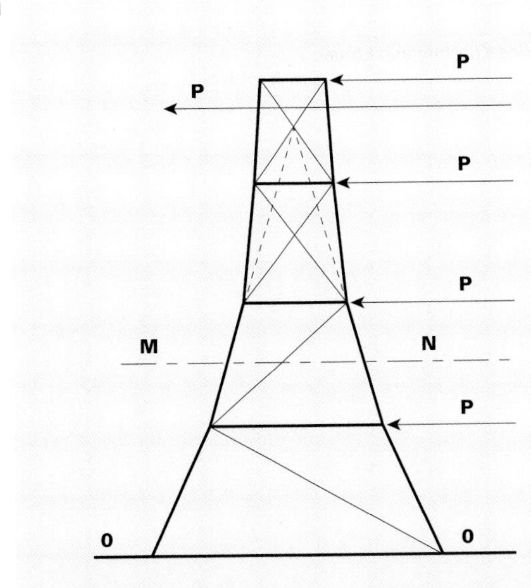

2

1
Sketch describing Eiffel's construction principle

2
Detail photograph of the Eiffel Tower showing the rivets

3
Overall view of the tower

3

4.2.4
Forth Rail Bridge

Structural description
Cantilever truss bridge

Location
Queensferry, near Edinburgh, Scotland

Completion date
1890

Length
2.5km

Engineers
Benjamin Baker (1840–1907), Allan Stewart and John Fowler

Contractor
Sir William Arrol & Co.

1

The Forth Rail Bridge, connecting Edinburgh with Fife, is the longest cantilever bridge in the world for rail transport, and the world's second longest such structure after the Quebec Bridge. It was designed by Benjamin Baker, Allan Stewart and John Fowler, who also oversaw the building work. The bridge was built by Glasgow-based company Sir William Arrol & Co. between 1883 and 1890, and was the first in Britain to be constructed using steel alone; up to this time, the strength and quality of steel yields could not be predicted.

The design concept for the bridge was illustrated by Baker in his 'human cantilever' model (see section 1.5). The bridge comprises two main spans of 521 metres with two spans of 207 metres at each end. Each of the main spans is made up of two cantilevering arms that support a 106-metre central truss. Connecting each end of the bridge to the river banks is a series of 51-metre span trusses.

The cantilever arms spring from three 100-metre-tall towers, which are built around four primary columns that each rest on a separate foundation. The southern group of foundations had to be constructed as caissons under compressed air to a depth of 27 metres. Whilst the two cantilevering arms that spring from each of the towers counterbalance each other, the shoreward ends carry weights of about 1,000 tonnes to counterbalance half the weight of the suspended spans and live load.

1
View of the Forth Rail Bridge under construction

2
The Forth Rail Bridge today

2

The use of cantilevers in bridge construction was not a new idea, but Baker's design included calculations for incidence of erection stresses, provisions for reducing future maintenance costs, calculations for wind pressures (evidenced by the Tay Bridge disaster) and the effect of temperature variation on the structure. A recent materials analysis of the bridge, ca. 2002, found that the steel in it – estimated to weigh between 54,000 and 68,000 tonnes – is still in good condition.

The weight limit for any train on the bridge is 1,422 tonnes, meaning that any current UK locomotive can use the bridge. Up to 200 trains per day crossed the bridge in 2006. The bridge is being considered for nomination as a UNESCO World Heritage site. During construction, over 450 workers were injured and 98 lost their lives.

4.2.5
All-Russia Exhibition 1896

Structural description
Hyperboloid tower;
steel, tensile enclosure;
double-curvature steel
gridshell

Location
Nizhny Novgorod, Russia

Engineer
Vladimir Shukhov
(1853–1939)

His 'gittermasts', attenuated hyperbolic paraboloids, were true minimum weight forms.[1]
Matthew Wells

The All-Russia Industrial and Art Exhibition of 1896 was held in Nizhny Novgorod on the left bank of the Oka River. The event was the biggest pre-revolution exhibition in the Russian Empire, and was organized with money allotted by Tsar Nicholas II. The All-Russia Industrial Conference was held concurrently with the exhibition, which showcased the best of Russian industrial developments from the latter part of the nineteenth century.

For the exhibition, the engineer and scientist Vladimir Shukhov pioneered the use of steel in a number of radical building types, including the world's first hyperboloid tower; the world's first steel, tensile enclosure; and the first double-curvature steel gridshell.

In the 1880s, Shukhov had begun designing roof systems that minimized the use of materials, time and labour. Probably based on Pafnuty Chebyshev's work on the theory of best approximations of functions, Shukhov invented a new system that was innovative both structurally and spatially; he derived a family of equations to enable the calculation and construction of hyperboloids of revolution and hyperbolic paraboloids.

Hyperbolic structures have a negative Gaussian curvature, meaning that they curve inward rather than outward. As doubly ruled surfaces, they can be made with a lattice of straight beams so remain relatively straightforward to build. Inspired by observing the action of a woven basket holding up a heavy weight, Shukhov solved the problem of designing lightweight, efficient water towers by employing a hyberbolic, steel, lattice shell. Owing to its lattice structure, the tower also experiences minimum wind load.

Shukhov called it *azhurnaia bashnia* ('lace tower'/'lattice tower'). The system was patented in 1899, and over the next 20 years he designed and built nearly 200 of these towers, no two exactly alike, with heights ranging between 12 and 68 metres.

1 Wells, M., *Engineers: A History of Engineering and Structural Design*, Oxford: Routledge, 2010, p. 130

1

1
The world's first double-curvature (diagonally framed) steel gridshell, shown during construction. The roofs of these pavilions were formed entirely of a lattice of straight angle iron and flat iron bars

2
The world's first steel, tensile enclosure – the Elliptical Pavilion of the All-Russia Exhibition, during construction in 1895

3
The Hyperboloid Water Tower – the world's first steel, lattice, shell structure, completed in 1896

4
Drawing of the 'gittermast'

5
Interior view of the mast looking up

6
View of the completed mast

2

3

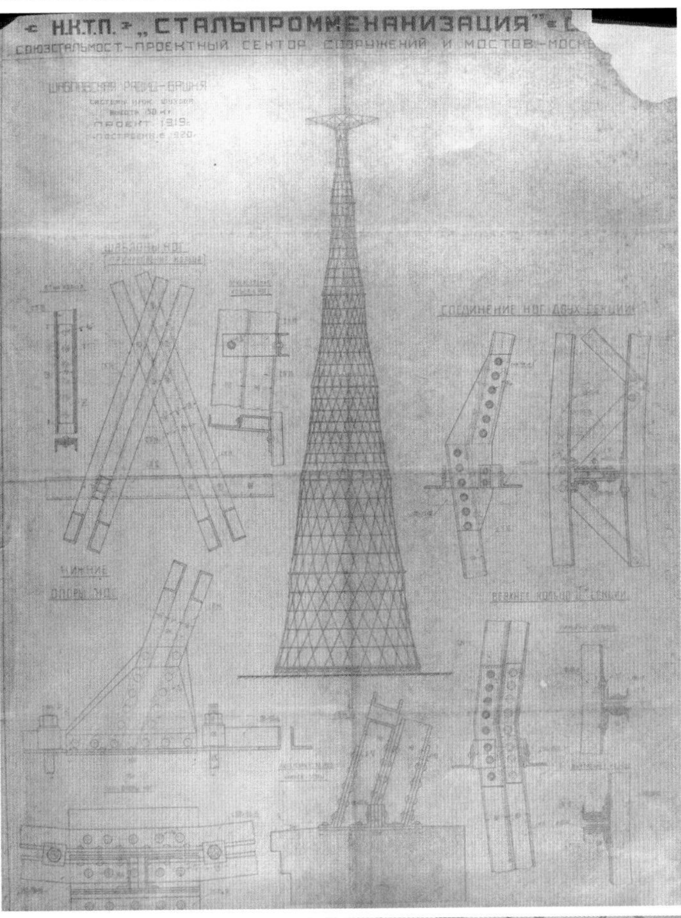

4

6

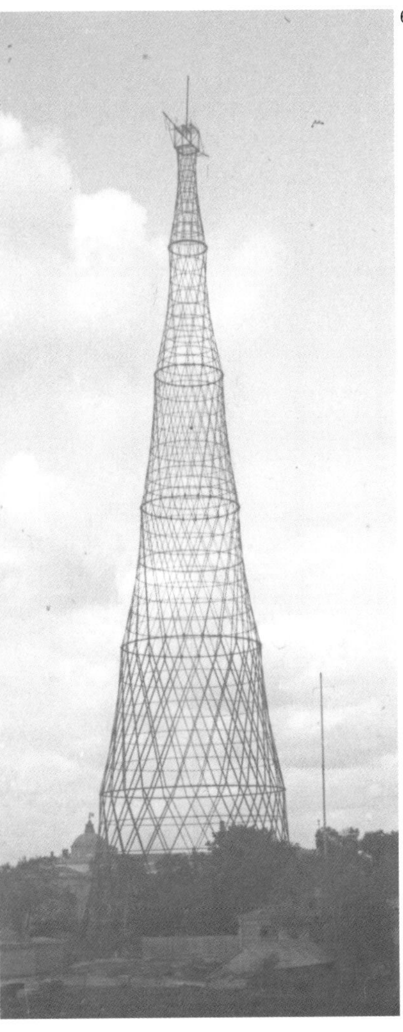

5

4.2.6 Tetrahedral Tower

Structural description
Octet truss spaceframe tower

Location
Beinn Bhreagh, Nova Scotia, Canada

Completion date
1907

Height
25m

Plan dimensions
Triangle with 1.8m sides

Designer
Dr. Alexander Graham Bell (1847–1922)

Engineer
Frederick Baldwin

Alexander Graham Bell discovered the octet truss while conducting research on flying machines.

Bell wanted to develop a kite that would be large enough to carry a man. In the same way that, in the second half of the century, the geodesic dome would solve Buckminster Fuller's problem of enclosing the maximum amount of space with the lightest structure, the tetrahedron enabled Bell to increase the size of a kite without increasing its weight. His first innovation was to remove a face from the standard box kite, producing a triangular section – lighter, more rigid and less prone to torsion under wind load. He went on to combine several small triangular kites, thus increasing the surface area with little increase in weight, until eventually settling on the tetrahedron, one of nature's most stable structures.

Bell appreciated that the kite structure might be applicable to ground-based, lightweight metal frames, and his experiments with tetrahedral cells culminated in the construction of an observation tower at Beinn Bhreagh, his summer estate near Baddeck, Nova Scotia.

Bell assigned the engineer Frederick Baldwin the job of building the tower, each 'cell' of which was composed of six 1.2-metre-long pieces of 15-millimetre-diameter ordinary galvanized iron pipe and four connecting nuts. Each cell could support 1,814 kilograms without stress. On completion in September 1907, the tower stood nearly 25 metres high.

The octet truss is now a common, standard component, used in many construction applications and seen every day in cranes throughout the world.

Dr. Bell said of his own, inventive ability to apply discoveries made in one field to another: 'We are all too much inclined, I think, to walk through life with our eyes shut. There are things all round us and right at our very feet that we have never seen, because we have never really looked.'[1]

1 Carson, M. K., *Alexander Graham Bell: Giving Voice to the World*, New York: Sterling, 2007, p. 118

1
Bell's design for a multi-celled, tetrahedral kite

2
Observation tower at Beinn Bhreagh, constructed using unskilled labour and sited deliberately so as to be subjected to high wind loads

1

2

4.2.7
Magazzini Generali Warehouse

Structural description
Reinforced-concrete, gabled, constant-force truss-supported roof

Location
Chiasso, Switzerland

Completion date
1924

Plan dimensions
33.4m wide x 25m long

Engineer
Robert Maillart (1872–1940)

1

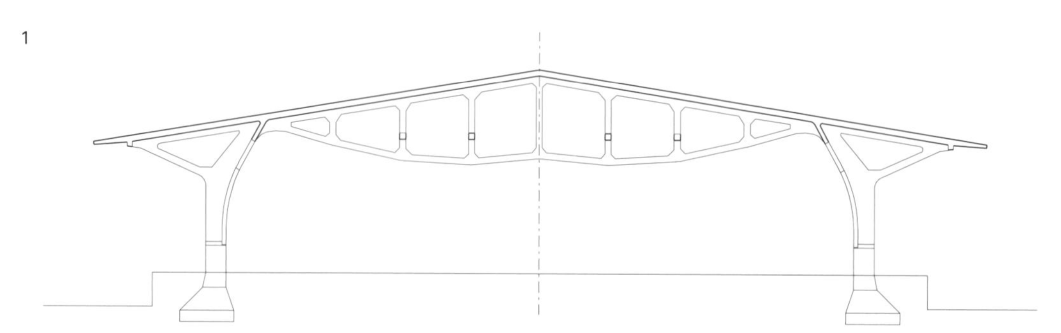

The truss takes up a plantlike form, reminiscent of certain structural forms of the 'art nouveau' period such as in ... the buildings of the Catalan, Antoni Gaudí.[1]
Max Bill

Built at the beginning of the twentieth century, this remarkable structure is still in use as a bonded warehouse for temporary storage of goods in transit at the Swiss/Italian industrial border town of Chiasso. It is an excellent example of a self-illustrating structural idea realized in *in situ* reinforced concrete. The most visually striking elements are the concrete trusses, which are cast in conjunction with the gabled roof slab. The thin roof slab acts to reinforce the compressive top chord of trusses that are supported by split 'tree-like' columns with cantilevered arms. The resultant form, given the dimensional constraints and a gabled cross-section for snow load, is an almost perfectly built diagram of evenly distributed internal forces.

The structure consists of six reinforced concrete trusses creating a clear span between structural supports of 25 metres. An additional covering 4 metres either side is created by a cantilevered edge, creating an overall covered width of 36.4 metres. The unique design means that all chords in the trusses are of the same cross-sectional dimension with the connecting elements between the top and bottom chord placed vertically, similar to a Vierendeel truss. In addition, the six trusses are also longitudinally connected by four linear elements to prevent buckling. The most formally complex elements of the structure are the 12 column supports, which bifurcate at the top with the major structural support pulled in on each side to pick up the truss, and with a smaller arm extending to the edge of the gable. These geometrically complex columns (T-shaped in section) are additionally shaped to reflect specific structural and functional requirements, including an enlarged, protected base for this storage depot and a longitudinal arched diaphragm wall where the column meets the truss, providing structural stiffening and an even load distribution.

In a paper for the Society for the History of Technology, authors Mark, Chiu and Abel[2] undertook a structural analysis of this unique building. Using numerical and photoelastic methods, they confirmed that its sculpted form is in fact more structurally expedient than is suggested by its carefully crafted appearance, wherein, as Max Bill stated in his monograph on Maillart, 'The form follows the flow of forces'. The results of their analysis showed 'almost uniform force levels' in the upper and lower truss chords, showing that the form is derived from, or at

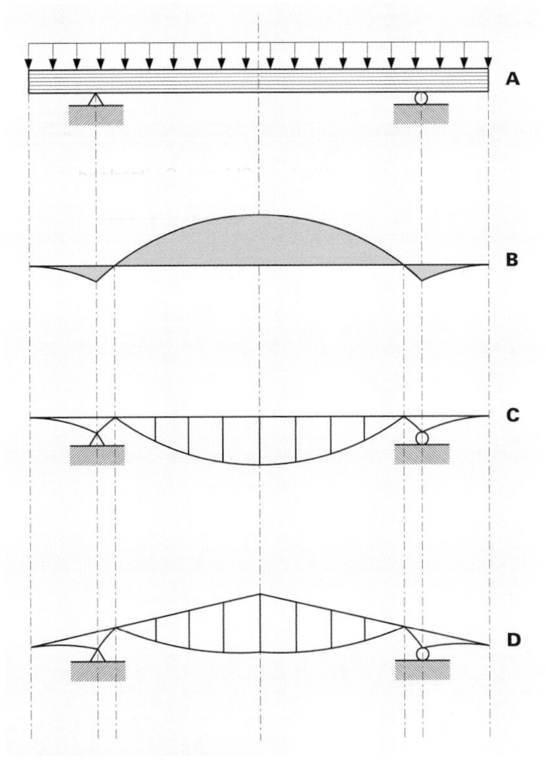

least closely replicates, the moment diagram. Analysis also showed that the monolithic construction, working in conjunction with the geometry of the cross-section and the designed connections, allows the roof to function as a type of 'stressed skin' structure. In conclusion, the analysis clearly confirms that the structural logic is successful and an even distribution of internal forces is achieved alongside specific programmatic and site requirements, specifically its industrial use and issues of snow loads.

Maillart's work with concrete was influential on the work of architects and engineers like Pier Luigi Nervi, who includes the Chiasso warehouse and shed in his book *Structures*, in the chapter 'Understanding Structures Intuitively'. The adjacent warehouse is also structurally interesting as an early example of flat-slab construction, wherein Maillart replaced beams in the structure with specially designed columns and column capitals designed to provide the necessary structural stiffness and axial support.

1
Cross-section drawing of Maillart's unique roof design

2
Main view of Chiasso 'shed' interior

3
Detail of cast column support at the roof edge

4
Detail of column and hanging truss connection

5
Diagram of the structural logic and development of the roof form:
A shows a simply supported beam
B shows the bending moment of that beam
C shows the reversal of that moment diagram
D is a diagram of Maillart's ultimate structural resolution of the Chiasso 'shed'

1 Bill, M., *Robert Maillart: Bridges and Constructions*, London: Pall Mall Press, 1969, p. 171
2 Mark, R., Chiu, J. K., and Abel, J. F., 'Stress Analysis of Historic Structures: Maillart's Warehouse at Chiasso' in *Technology and Culture*, Vol. 15, No. 1 (Jan. 1974), pp. 49–63

4.2.8
Zarzuela Hippodrome

Structural description
Doubly curved reinforced concrete shell structures

Location
Madrid, Spain

Engineer
Eduardo Torroja (1899–1961)

Completion date
1935

Eduardo Torroja, the Spanish engineer, was born in 1899 into a family of mathematicians, engineers and physicists. He was the founder of the International Association for Shell Structures (IASS) and, at its peak in the 1930s, Eduardo Torroja's Engineering Bureau was producing many innovative designs and experimental construction techniques, including early developments in prestressing concrete.

In 1959, at a time when shell structures were frequently used all over the world to roof buildings such as sports and exhibition halls, industrial plants, silos and cooling towers, Torroja organized and convened an International Colloquium in Madrid. During this colloquium, Torroja proposed the founding of the IASS.

Torroja is quoted as saying: '...far more than the technical results, I value the experience in its human, social and professional dimension ... to create organizations where the different professions, the upper and lower echelons, could work together in perfect harmony; where everyone has grown accustomed to living a life on the highest rung of humanity, where courtesy, mutual respect and support, and maximum personal dignity reign.'[1]

1 Schaeffer, R. E., *Eduardo Torroja: Works and Projects*; book reviewed by Pilar Chías Navarro and Tomás Abad Balboa in *Journal of the International Association for Shell and Spatial Structures* (IASS), Vol. 47, No. 3, December, 2006, p. 152

1

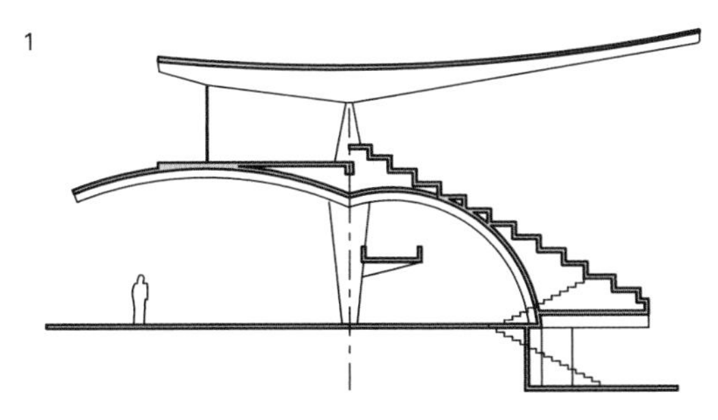

2

3

1
Section through the roof

2
Aerial view showing the double-curved roof under construction

3
The roof of the grandstand at the Zarzuela racecourse cantilevers some 13m

4.3
1950–1999

4.3.1
Crown Hall, Illinois Institute of Technology (IIT)

Structural description
Steel portal frame with cantilevered ends

Location
Chicago, USA

Completion date
1956

Plan dimensions
67m long x 36.6m wide

Height
8.36m

Architect
Ludwig Mies van der Rohe
(1886–1969)

1

2

Where technology attains its true fulfilment, it transcends into architecture.[1]
Ludwig Mies van der Rohe

One of Mies van der Rohe's most celebrated works, Crown Hall remains an elegant and concisely engineered structure well over 50 years after its completion. Built as part of a 48-hectare campus entirely designed by Mies in the Bronzeville neighbourhood of Chicago, Crown Hall remains the centrepiece of this remarkable architectural park, which is still the main IIT campus out of five that the institute has in the city. Crown Hall was designed to house the faculty of architecture and town planning (a very deliberate, proximate relationship), and Mies had a particular interest in this project as he directed the architecture program at IIT from 1938 until 1958. The building is arranged over two levels and uses a planning module of 3.05 metres. To enter, you ascend 1.8 metres on travertine stone steps and enter the clear-span space of the main 'studio' floor, a single-volume space 67 metres long, 36.6 metres wide and 5.48 metres between terrazzo floor and white-painted acoustic ceiling. Two stairs to the lower floor – leading to additional lecture, teaching and library spaces – punctuate the largely unobstructed ground-floor level. The main floor also contains low, freestanding oak-clad partitions and two non-structural, slim service risers, which are the only floor-to-ceiling elements.

While there are no gratuitous structural gymnastics, Mies cleverly reverses a typical beam-and-roof arrangement and sets the four main structural beams at 18.3-metre centres across the outside of the roof, supported by eight external columns, forming welded portal frames made from hot-rolled steel sections. This structural arrangement maintains a perfectly clear space and smooth uninterrupted soffit. The nature of the fabricated steel construction also creates usefully rigid connections and obviates the need for any visible cross-bracing. The roof projects 6.1 metres beyond

1
Main entrance to Crown Hall, with two of the four fabricated plated beams

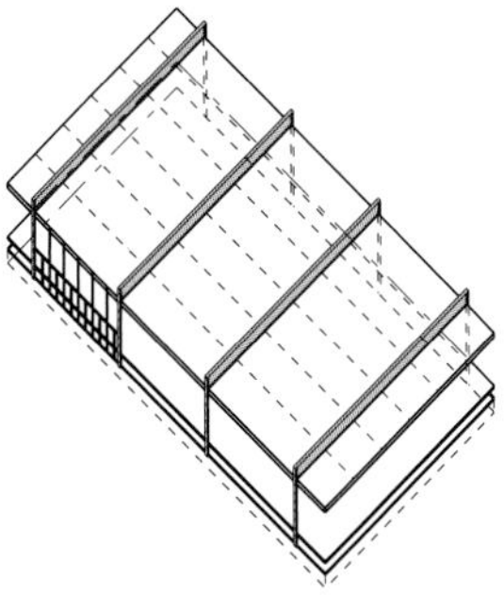

2
Rear entrance to Crown Hall

3
Axonometric illustration showing the structural assembly

4
Detail of column support and plate-steel support beam, which incorporates an access ladder of square-section steel welded to the column flanges

5
Corner detail, showing sandblasted glass at lower level

the main steel frames at each end, enhancing the effect of its underside as a kind of floating plane. This steel-framed prism is glazed on all sides with sandblasted (translucent) glazing to the lower panels. The building was renovated in 2005 by Kreuck & Sexton Architects, who undertook a thorough and fastidious job that involved an entire reglazing, sandblasting the steelwork and repainting in an appropriate 'Mies Black' that did not contain lead and will not fade in sunlight.

The steelwork for Crown Hall is a mixture of standard hot-rolled column and angle sections, and specially fabricated steel components. Eight larger columns at 18.3-metre centres support the four custom-fabricated plate girders, which are 1.83 metres high. Intermediate, smaller H-section columns are located at 3.05-metre centres and delineate the glazing regime. With all structural elements visible and clearly expressed, and featuring impeccable detailed design, Crown Hall is still an exemplary model of well-worked structural logic, elegant material composition and forthright utility. Mies had also used the 'exterior structure' logic found at Crown Hall in his National Theatre, Mannheim competition entry of 1953, although for this much larger structure, 160 metres long and 80 metres wide, he had proposed to replace the solid steel of the plate girders with an open steel lattice truss. Mies executed many projects in Chicago, but alongside his Lakeshore Drive apartments, the newly restored Crown Hall remains one of his most potent and enduring works.

1 Blaser, W., *Mies van der Rohe*, London: Thames and Hudson, 1972, p. 80

4.3.2 Los Manantiales Restaurant

Structural description
Reinforced-concrete hyperbolic shell

Location
Xochimilco, Mexico City, Mexico

Completion date
1958

Engineer
Félix Candela (1910–1997)

Architects
Fernando Alvarez Ordóñez, Joaquin Alvarez Ordóñez

It may be said there are two basic criteria for a proper shell: the shell must be stable and of a shape which permits an easy way to work. It should be as symmetrical as possible because this simplifies its behaviour. Either interior groins (as in the restaurant in Xochimilco) or exterior edges should be able to send loads to points of support, or else there should be a continuous support along certain edges.[1]
Félix Candela

Candela collaborated with Colin Faber on the general form of the restaurant. The form of the shell is a 'groined' vault, made up from four intersecting hyperbolic parabolas with curved edges free of any edge stiffeners so as to reveal the thinness of the shell. The groins are the valleys in the shell, formed at the convergence of the intersecting hyperbolic parabolas.

Candela stiffened the groins using V-section beams. These V-beams are reinforced with steel, while the rest of the shell has only nominal reinforcing to resist local cracking. For the foundations, Candela anchored the V-beams into footings shaped like inverted umbrellas to prevent the shell from sinking into the soft soil. The footings were then linked with steel tie bars to resist lateral thrusts from the shell.

Hyperbolic parabolas may also be understood as ruled surfaces. That is to say, their three-dimensional geometry can be generated by series of straight lines. The form boards for construction followed the path of these straight-line generators. Once the reinforcing steel mesh had been laid on them, the concrete was poured by hand, one bucket at a time.

1 Faber, C., *Candela: The Shell Builder*, New York: Reinhold Publishing Corp., 1963, p. 199

1

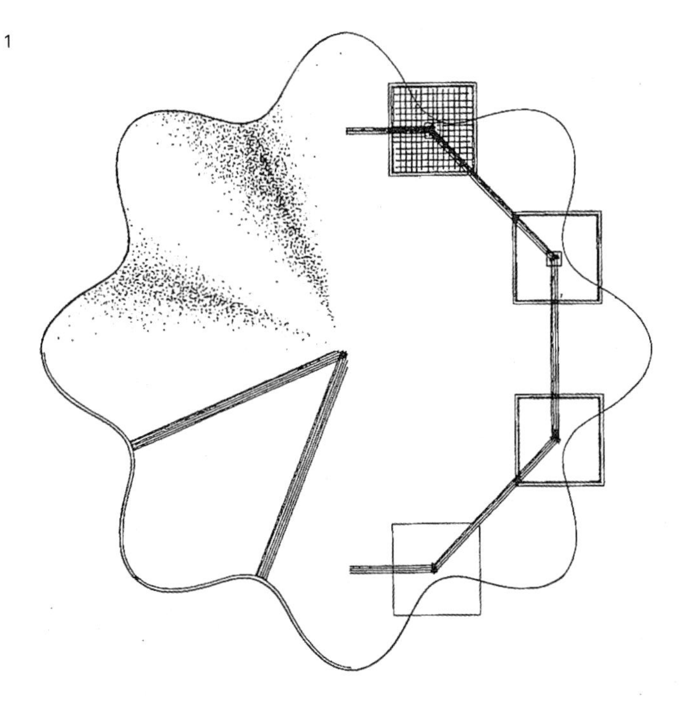

1
Plan

2
Section and elevation

3
The building during construction

4–6
The building today

2

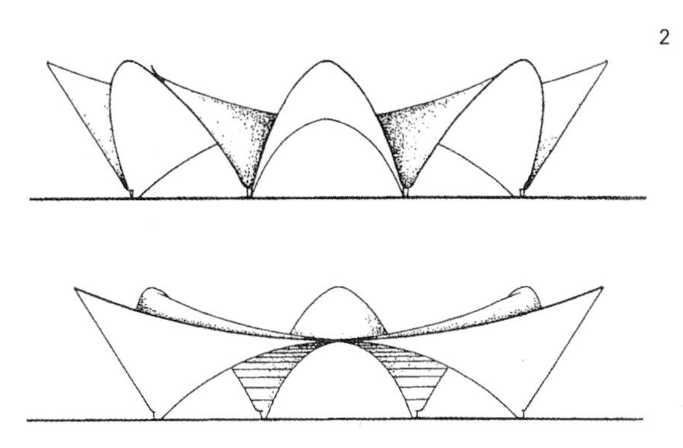

3

5

4

6

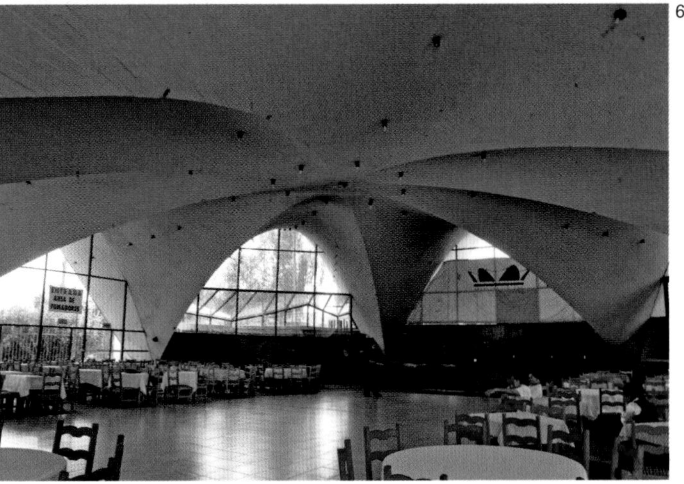

4.3.3
Concrete Shell Structures, England

Structural description
Concrete shell structures

Location
Markham Moor and Ermine, Lincolnshire, England

Architect
Sam Scorer (1923–2003)

Completion dates
1959/1963

Sam Scorer was a prolific architect, and in addition to his pioneering work on shell structures he carried out much building-conservation work including major restoration on Lincoln Cathedral. He was chairman of his local planning committee, produced *Architecture East Midlands* magazine in the mid-1960s and was a Fellow of the Royal Society of Arts.

As a talented painter in his own right, and in giving vent to his artistic frustrations, Scorer came up with radical designs for hyperbolic-paraboloid (doubly curved) roofs – most notably in a Lincoln church and what is now a roadside restaurant on the A1 at Markham Moor.

The last-named was originally designed as a canopy over a petrol station, extending at one end from a long, low building housing an office and kiosk over a row of pumps. A few years after its construction in the late 1950s, a restaurant was built underneath the flying roof. Early in the new millennium it was threatened with demolition to make way for a slip road, but a campaign in 2005 granted it a reprieve.

Unlike the restaurant, the interior of St. John's Church fully benefits from having such a majestic roof – its saucer shape effectively eliminating the need for columns, allowing a large interior space unencumbered by structure. From the outset, Scorer was interested in how theology related to the building, what the church stood for, how it worked and how it related to the community, the emphasis very strongly being one of a 'tent of meeting'. The font is at the lowest point of the church and the altar, also designed by Scorer, at the highest – all presided over by a fine stained-glass window designed by Keith New, who also designed windows in the rebuilt Coventry Cathedral.

The pouring of the 75-millimetre-thick concrete roof at St. John's had to be done in one continuous operation, in very frosty conditions. Although kerosene burners were employed to prevent freezing, hairline cracks appeared on drying, requiring additional support for the concrete tie beam beneath the floor, for which more concrete was added to the two pools of water (that reflect the significance of baptism) at either end of it. The outer surface of the roof was covered in fibreboard and super-purity aluminium (since re-clad in a proprietary membrane in the late 1990s, after damage). The shuttering of the roof had such a fine appearance that it was retained, producing a fine ceiling comprising a mass of timber slats. The church was completed in 1962 and was listed in 1995.

1

5

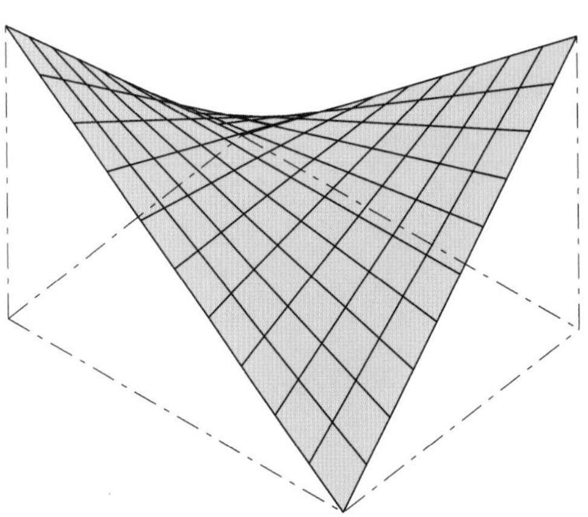

1
Petrol station canopy, Markham Moor

2–4
St. John's Church

5
Diagram of the hyperbolic-paraboloid roof structure

6
Copy of one of Sam Scorer's sketches for the church bell tower

7–8
St. John's Church

2

6

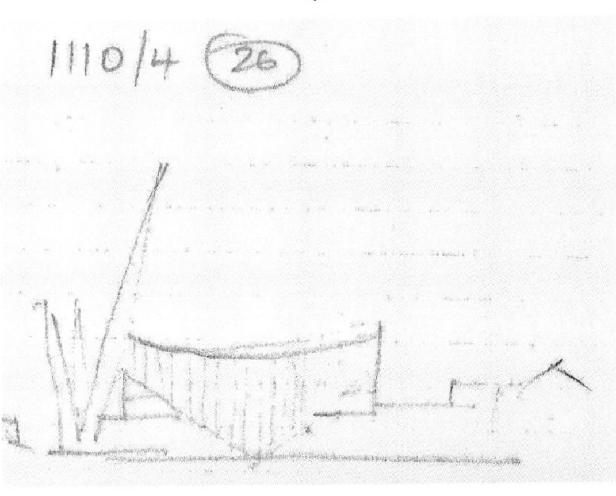

3

7

4

8

4.3.4 Geodesic Domes

Structural description
Geodesic dome structures

System designer
Richard Buckminster Fuller
(1895–1983)

...world engineering not only was surprised by the geodesic behavior but clearly stated that it was unable to explain or predict the unprecedented performance per pound efficacy of the geodesic structures by any of the academically known principles of analysis.[1]
Richard Buckminster Fuller

When discussing the work of Richard Buckminster Fuller (also referred to as 'Bucky'), it is difficult to do so without mentioning his larger theoretical and philosophical project for what he called 'Spaceship Earth'. This project, which lasted the duration of his professional life, encompassed a highly tuned, environmental, 'humane' consciousness and interests that were to include energy, transportation and servicing systems with special attention given to that most fundamental of human needs, shelter. As an ex-US Navy man, Fuller recognized the logistical and operational excellence of this highly resourced organization, if not its socio-political *raison d'être*. Buckminster Fuller declared that the designer must concern himself with 'Livingry' not weaponry, and so began his lifelong experiment 'of what one man can do', which was to embrace art, architecture, engineering and poetry. As well as being a highly skilled and articulate strategist, Fuller also interested himself in what he described as the artefacts of his ideas, which in themselves are highly original. These inventions include several patented structural systems, most notably the geodesic dome, which latterly became synonymous with Fuller. It is worth noting that while geodesic geometry and geodesic domes are an end (or artefact) in themselves, they are also closely related to Fuller's social and technological mapping of the world, with the mathematics of geodesy crucial to establishing networks of food distribution, energy systems, freshwater supply and shelter. As his long-time collaborator Shoji Sadao recently noted, 'For Bucky, the problem of transferring the planet's spherical form on to a two-dimensional piece of paper had not been resolved satisfactorily.'[2]

The geodesic dome patented by Fuller in 1954 is known to be the most structurally efficient of the domes derived from the icosahedron (a 20-sided polyhedron). In the patent application, Fuller described it as a spherical mast, which evenly distributes tension and compression throughout the structure. The form combines the structural advantages of the sphere (which encloses the most space within the least surface, and is strongest against internal pressure) with those of the tetrahedron (which encloses the least space with the most surface and has the greatest stiffness against external pressure). A geodesic structure distributes loads evenly across its surface and, as with a spaceframe, is efficient to construct, as it is composed entirely of small elements. The geodesic dome is the product of a geometry based on the shortest line between two points on a

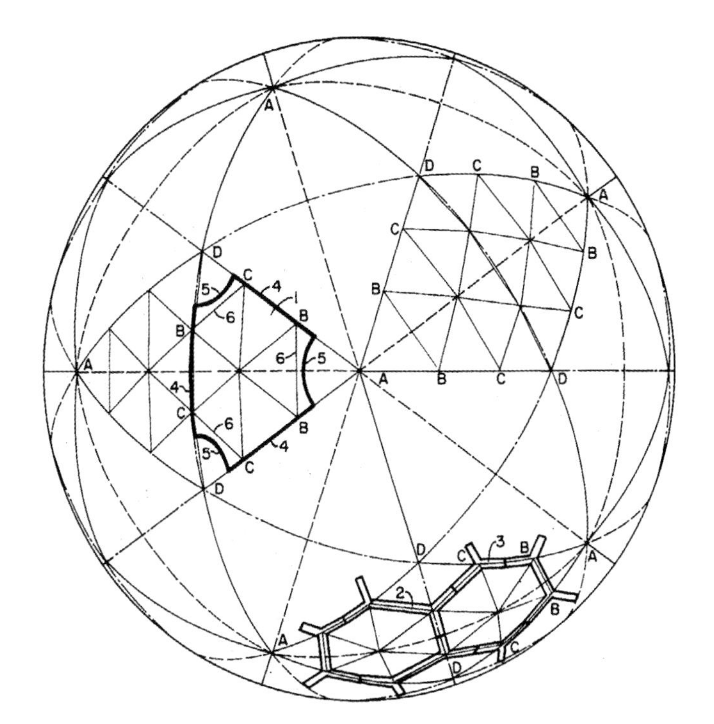

Figure 1 from Fuller's US Patent 3,197,927, in which he describes different geodesic structural configurations based on the 'great circle' subdivision of a sphere

mathematically defined surface; it takes its name from the science of geodesy – measuring the size and shape of the Earth. It consists of a grid of polygons that is the result of the geodesic lines intersecting. The number of times that you subdivide one of the triangular icosahedra faces is described as the 'frequency'; the higher the frequency, the more triangles there are, and the stronger the dome will be. The scalability of the geodesic dome is interesting, with Fuller pointing out that '... every time a geodesic dome's diameter is doubled, it has eight times as many contained molecules of atmosphere but only four times as much enclosing shell...'.[3] This realization led to Fuller's proposal in 1950 to enclose the whole of midtown Manhattan in a 3.2-kilometre-diameter geodesic dome, whose enclosure would have weighed significantly less than the volume of air contained within and whose structure would be largely rendered invisible because of physical proximity and our relative visual acuity.

Fuller and his consultancy companies, Synergetics and Geodesics Inc., produced many structural types of geodesic enclosure, working in collaboration with other architects and engineers. Fuller also licensed his technology, which comprised the patented geometric configuration and various connection details. Domes were fabricated from a wide range of materials, which included cardboard, plywood sheets, sheet steel and fibre-reinforced plastics. Four key domes and dome types are described overleaf. They utilize different materials and fabrication processes but are all derived from Fuller's geodesic geometry.

1 Krausse, J., and Lichtenstein, C., *Your Private Sky: R. Buckminster Fuller*, Zürich, Lars Müller Publishers, 2001, p. 229
2 Sadao, S., *A Brief History of Geodesic Domes, Buckminster Fuller 1895–1983*, Madrid: AV Monographs 143, 2010, p. 87
3 Fuller, R. B., *Critical Path*, New York: St. Martin's Press, 1981, p. 209

The Climatron, St. Louis, Missouri, 1960

Architect
Murphy and Mackey Architects

Plan dimensions
53m diameter

Height
21m

This climate-controlled enclosure has recently celebrated its 50-year anniversary as a tropical and subtropical greenhouse at the Missouri Botanical Garden. The clients had wanted a large space without any internal walls or supports, which led them to Fuller's new technology. The structure is a quarter-sphere dome fabricated from tubular aluminium sections, acting in compression, which are bolted to cast connector joints (or nodes) with tensile forces carried through interconnected aluminium rods. The structure is held aloft on a series of structural-steel articulated columns. The Climatron was originally clad in acrylic plexiglas panels, which were replaced with glass and an additional support frame in the 1990 refurbishment.

1
Recent photograph of the restored Climatron dome

2
Detail of the Climatron's aluminium structural frame. Note the reciprocating tension rods

Wood River Dome, Wood River, Illinois, 1960

Architect
R. Buckminster Fuller with Synergetics Inc (James Fitzgibbon and Pete Barnwell)

Plan dimensions
117m diameter

Height
36m

This dome is the less-celebrated near relation of The Union Tank Car Building in Baton Rouge, Louisiana, which was demolished in 2008. When the Baton Rouge dome was constructed in 1958, it was the world's largest clear span – a record that it held for 11 years. The Wood River dome is an almost identical construction, albeit with a less elaborate interior. Both structures are geodesic exoskeletons of welded tubular steel, fixed to a folded 2.78-millimetre (12-gauge) welded sheet-steel skin, which acts in tension as well as providing the environmental envelope. The Wood River dome was built from the top down, with the structure gradually raised pneumatically with a huge air-inflated fabric bag. The building remains in use for the servicing of railcars.

3
Recent picture of the Wood River dome

4
Detail of the Wood River dome showing the sheet-steel structural skin and tubular-steel exoskeleton

Geodestic (Fly's Eye) Dome, Snowmass, Colorado, 1965

Architect
R. Buckminster Fuller

Plan dimensions
8m diameter

Height
6m

In 1965, Fuller was granted a patent on his Monohex-Geodestic structures, which he also called the Fly's Eye Dome. While still based on his geodesic geometry, he utilized the configuration of pentagons and hexagons that we recognize as a simple football. Fuller then created holes (the 'eyes') in the pentagons and hexagons, leaving a triangular-shaped component to connect them. Combined with this geometric development, Fuller also used the plastic properties of Glass Reinforced Plastic (GRP) to create an additional upstand around the openings. This compound (or double) curvature creates a very strong construction component.

5
An 8-metre diameter Fly's Eye Dome, made from 50 GRP panels, which are bolted together using 2,000 stainless-steel bolts

The USA Pavilion, Montreal, Canada, 1967

Architect
R. Buckminster Fuller, Shoji Sadao

Plan dimensions
76m diameter

Height
61m

The pavilion was constructed for the Montreal Expo of 1967, and consisted of a three-quarter sphere, geodesic, double-layered, tubular-steel space grid. Fuller's geodesic dome was originally weathered using 1,900 moulded acrylic panels, which incorporated six triangular sun blinds within each six-sided panel, and were automatically opened or closed in response to the movement of the sun in relation to the structure. This remarkable structure still exists as an ecological museum overlooking the city of Montreal. Interestingly, if you look for the equator (or the horizontal half-point) of the dome you will see that the horizontals below (towards the ground) are parallel and of decreasing circumference, whereas the structural geometry above the equator is purely geodesic.

6
Recent composite photograph of the Montreal dome

7
Detail of the Montreal dome showing welded-steel tubular framework

4.3.5
Palazzo del Lavoro (Palace of Labour)

Structural description
Reinforced-concrete and steel cantilevered canopies

Location
Turin, Italy

Completion date
1961

Floor area
45,000m² (25,000m² on the ground floor)

Height of canopies
20m

Architects
Pier Luigi Nervi (1891–1979) and Antonio Nervi

Contractor
Nervi & Bartoli

Pier Luigi Nervi was one of the great architect/engineer/builders of the latter part of the twentieth century. He worked as a structural engineer and designer on some remarkable collaborations, such as Giò Ponti's Pirelli Tower (Turin 1955–6) and with Marcel Breuer on the UNESCO Headquarters (Paris, 1955–8). It is, however, his single-authored works or collaborations with his son Antonio that remain his most distinct contributions to the field of architecture and construction. The Palazzo del Lavoro is one such project; it is particularly notable for the speed of its construction, which provided 45,000 square metres of exhibition space in little over 11 months. The unusual structural design, comprising 16 independent column/canopies, was employed in order that finishing work could run concurrently with structural work – a criterion that would have proved problematic with a single roof structure.

Each roof structure, or 'mushroom' canopy, measures 40 metres square and 20 metres high. The mushroom support columns are cast in six vertical sections, with the steel formwork for each section subdivided into four reusable moulds that were bolted together. The casting of each column lasted ten days. The horizontal joint lines are clearly detailed as a recessed shadow gap. The geometric form of each column transforms from a 5-metre-wide cruciform at the base, to a circular profile at the top, with a 2.5-metre diameter. The surface finish of the columns is further articulated by the close vertical board markings of the inside of the shuttering mould. Originally the mushroom canopy structures were conceived in concrete, but for speed of construction the canopy elements were prefabricated off-site as a series of 20 tapered steel fins radially arrayed around a central hub, which is bolted to the concrete column head. A gap of 2 metres is left between each structural canopy and a glazed element introduced in order to provide rooflights – with the external glazed envelope created by 'Jean Prouvé type' folded-steel mullions, which are held on hinged connections to allow for thermal expansion. In addition to the main internal exhibition floor, a mezzanine level wraps around three sides, independently supported by a separate column grid and with an *in situ* cast slab featuring Nervi's innovative 'isostatic' rib geometries.

Nervi, writing in his book *Structures*, explains that his employee, Aldo Arcangeli, had suggested that the ribs of a concrete slab should follow the lines of a structure's principal bending moments. These isostatic lines were made visible in the technology, relatively new for the period, of photoelastic modelling, wherein the stress patterns of a clear substrate are made visible through polarized light. By constructing a scale model of a structure in a clear epoxy, Nervi began to create a new development of the surface geometry and structural behaviour of a concrete floor slab. He first employed this new technique in the Gatti Wool Mill in Rome (1951), where 16 curved ribs connect back to each column head in a repeated pattern, which is beautifully reproduced using reusable ferrocement formwork – another technological development pioneered by Nervi. However, structural engineer Matthew Wells, writing in *Engineers: A History of Engineering and Structural Design*, is broadly dismissive of these 'isostatic' lines as

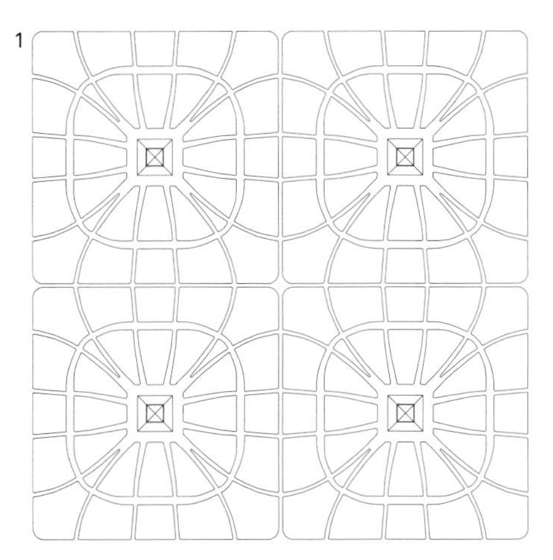

1
Reflected ceiling plan, showing the isostatic rib layout prototyped at the Gatti Wool Mill, Rome (1951)

meaningless and paradoxical, in that by reflecting structural action in built form you thus affect the structural action that you had originally modelled. Given Nervi's role as a designer these observations seem petty, as structural optimization was perhaps only one of many factors influencing the conception, engineering and construction of his work. It is worth noting that Nervi was a builder as well as an architect and engineer, and that it seems unlikely that without his direct involvement – and that of his builder cousin, John Bartoli – he would have been able to produce such structurally and architecturally ambitious forms.

Pier Luigi Nervi also made a considerable contribution to the construction industry through new production processes, including prefabrication and material innovations such as ferrocement, which was developed and patented by Nervi and Bartoli as a new construction material. Ferrocement consisted of the use of a strong cementitious mortar mix, built up over densely packed fine metal-mesh reinforcement. Originally pioneering it for shipbuilding in the 1940s, Nervi was determined to develop a reinforced-concrete technology that could dispense with labour-intensive timber formwork for complex geometric forms and that simultaneously optimized the structural performance of the material, creating what Claudio Greco calls 'a more homogenous and efficient composite'.[1] Working in conjunction with Professor Oberti at the Politecnico of Milan, Nervi's tests on the ferrocement revealed a considerable improvement of the tensile strength of the material in comparison with ordinary reinforced concrete and its relatively crude distribution of tensile steel reinforcement. The fabrication process of Nervi's ferrocement also meant that expensive and complex timber formwork could be largely dispensed with, as the fine steel meshes, densely layered into a fibrous matrix, could hold their shape whilst a cement mortar is hand-applied with trowels. Ferrocement was used for the highly detailed, reusable 'isostatic' formwork moulds and in its own right as a thin, cementitious panel. Notable ferrocement projects include the prototype Nervi- and Bartoli-designed storage building, Magliana (Rome, 1945), fabricated in undulating panels of 30-millimetre-thick ferrocement; and the *La Giuseppa* motorboat, constructed in 1972 and still seaworthy almost 40 years later. Nervi and Bartoli's skilled workforce was also used in the prefabrication of building components. Using the processes and techniques of the terrazzo and concrete industries, which worked in both prefabrication and *in situ* cases, Nervi was able to control quality, programme and cost. Interestingly, in the Palazzetto dello Sport (Rome, 1957) he employed a combination of pre-cast trapezoidal concrete panels (variously sized, with protruding steel reinforcement) with *in situ* concrete ribs cast between them, forming downstand beams to ensure a structurally integral whole.

1 Greco, C., 'The "Ferro-Cemento" of Pier Luigi Nervi, The New Material and the First Experimental Building' in *Spatial Structures: Heritage, Present and Future, proceedings of the IASS International symposium 1995, June 5–9, 1995*, Milan: S.G.E. Publishers, 1995, pp. 309–316

2

3

4

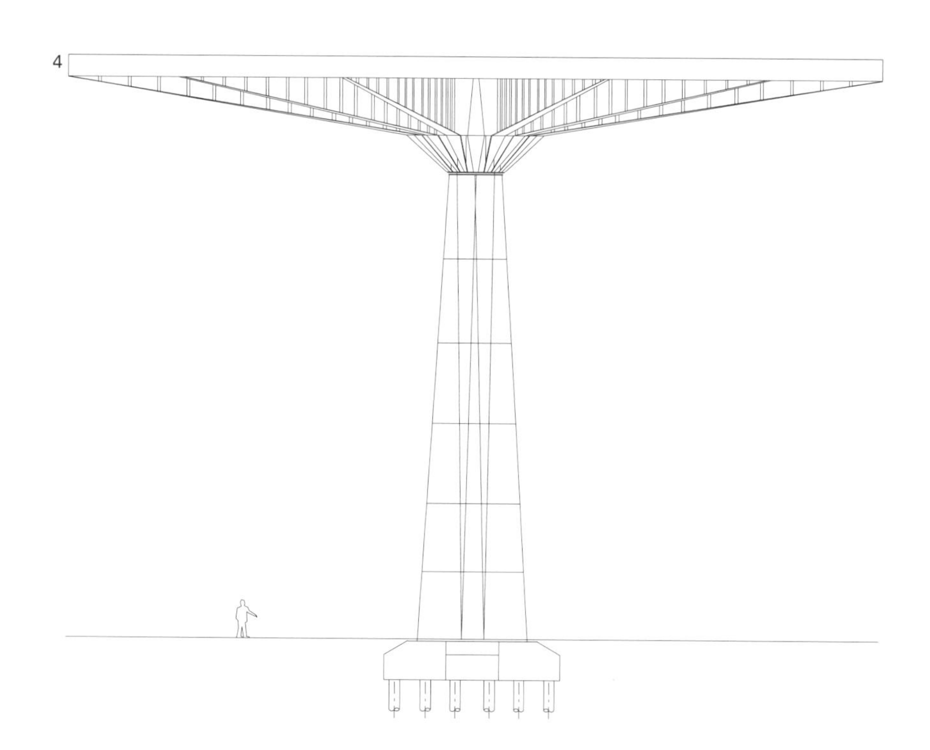

5

6

2
Recent interior view of the Palazzo del Lavoro, showing the independent 'mushroom' canopies

3
Detail of column form and the transition from a cruciform base to a circular column head

4
Elevation of one of the 16 'mushroom' canopies

5
Detail of column head and radial steel fins

6
Detail of canopy soffit

4.3.6
Concrete Shell Structures, Switzerland

Structural description
Reinforced concrete shells

Location
Wyss Garden Centre/
Deitingen Süd Service
Station/Brühl Sports Centre
Solothurn, Switzerland

Completion date
1962/1968/1982

System designer
Heinz Isler (1926–2009)

Contractor
Willi Bösiger AG

Over a period of more than 40 years, Swiss-born engineer Heinz Isler created a unique body of work. His material was reinforced concrete, with which he created a built encyclopaedia of thin concrete-shell structures, through a process of intuitive form finding coupled with modelmaking, prototyping and analytical tools of his own devising. His work is primarily located in Switzerland, but the quality of the projects and his unique working methods have extended his influence much further afield.

What is striking when visiting Isler's Swiss projects is how well maintained they are – with the exception of the service-station roof at Deitingen, which is structurally intact but with which its international oil company tenant's corporate identity does not chime. His factory buildings, sports and garden centres are clearly coveted by their enlightened owners, who recognize their interesting fusion of structural and material efficiency with a highly expressive form. As self-illustrating structural concepts, these delicately frozen membranes absolutely confirm the structural integrity of specific geometric forms, material properties, gravity and scale.

Isler's concrete roof shells can be crudely divided into three main types: bubble shells, free-form shells and inverted membrane shells. Bubble shells were one of his first real structural innovations in the development of large-span shells. Inspired by the geometry of a pillow, Isler developed a test rig where he could inflate a rubber membrane to form a double-curved synclastic 'pillow' shape in pure tension, which logic suggested would form a compression shell if inverted. Isler's testing included coating the inflated structure with plaster and then accurately measuring the curvature of the surface, with his own measuring jig, a calibrated pointed steel rod relocatable and capable of measuring to within one-fiftieth of a millimetre in the x, y and z dimensions. When challenged about the consistency of this empirical approach to structural development and testing, Isler described how the measured data would be plotted as a series of two-dimensional curved profiles, with any inaccurate measurements clearly showing. During structural modelling of the square-plan bubble shells, Isler was surprised to find that the static load of the structures was not evenly distributed to the four edges of the shell, but that 90 per cent of the total load was distributed to the four corners. This discovery has subsequently seen the bubble shells employed for literally hundreds of mostly industrial projects for large factory, warehouse and transport purposes. The shells typically range from 15- to 40-metre spans, feature a circular opening at their apex for daylighting and ventilation, and are clad with a fibre-reinforced plastic dome, also developed by Isler. In profile, these structures feature an edge beam that doubles as a gutter and has a span-to-depth ratio of 1:25; the circular openings are reinforced by an upstand approximately 250 millimetres deep, although the main structural shell is only 80 to 100 millimetres thick.

The other key Isler shell type that utilizes form-finding techniques is the inverted membrane shell, where a hanging membrane or flexible grid is hung from four corners, loaded and subjected to gravity. The resultant tensile form is then made rigid and turned upside down to form a self-supporting

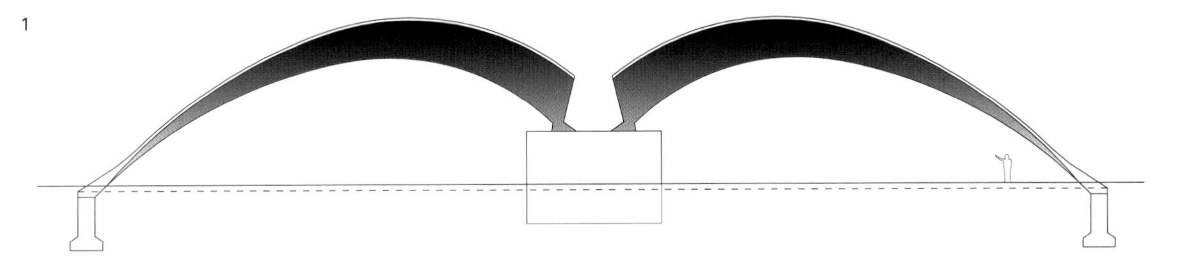

1
A cross-section of the inverted membrane shell of Deitingen Süd Service Station

compressive structure. Isler used many modelling techniques to create these forms, including fabric saturated in wet plaster or resin that was then allowed to dry before inverting the surface to create a prototype structure. Isler also discovered other useful structural devices in the form-finding process – and that by hanging a fabric membrane from four points set in from the corners, the free-hanging edge material forms a beam or arch structure when rigidized and inverted. A key example of the inverted membrane shell technique is the iconic Deitingen Süd Service Station project, where two identical triangular (in plan) three-point-supported shells, each 26 metres wide, span 32 metres with a pure compressive steel-reinforced concrete shell of only 90 millimetres' thickness. The relationship between the support points of such structures is important to note, and to avoid hugely costly slab foundations the support points are literally connected with prestressed tension ties.

The third key type of Heinz Isler shell structures comprises what he calls free-form shells. These are not derived from the form finding of inflation or hanging-gravity catenary models – or by mathematics, such as the 'anticlastic', or saddle-shaped, form of a hyperbolic paraboloid – but through a graphic process of carefully interfaced radii and compound curves. The garden centre pavilion at Wyss is an early example of such a structure, from 1962. With a span of 24 metres, a shell only 70 millimetres thick is created, which has four support points. The original curtain-wall glazing for these buildings was hung from a series of slender prestressed mullions. The free edges of the shell are turned up, to form a kind of stiffened arch between supports and to direct rainwater to the corners. The external surface of the Wyss shell was, and is, painted, whereas most of Isler's shells are not. This was primarily an aesthetic decision, but also a recognition that this type of shell is not a purely compressive structure and that where areas of tension occur local cracking might appear, making the structure susceptible to rainwater. The building is almost 50 years old, and although the glazing system has been refurbished the shell remains in excellent condition.

The importance of continual modelling and testing was key to the success of Isler's projects, as was a highly skilled construction team. The fabrication of a concrete shell requires a large amount of timber formwork and attendant carpentry – a fact of which Isler was well aware. In order to mitigate waste, he began to utilize woodwool panels as a permanent shuttering and interior finish, which was both thermally and acoustically beneficial. He also designed reusable glued laminated (glulam) timber formwork for products such as the bubble shells.

One of the key features of Isler's work is that the process of design, engineering and construction of these shell structures is all under his close control, and that only the process of modelmaking and prototyping (sometimes at full scale) would have allowed the construction of these unique projects. Surely the most elegant illustration of his ideas, and certainly the most ephemeral, are the ice forms he constructed by hanging fabric, which he then saturated with water before the Swiss winter completed the process, forming delicate ice shells.

2
Wyss Garden Centre shell, almost 50 years after its construction

3
Detail of Wyss shell, showing the cantilevered 'folded' edges that protrude at the central span by 3.5 metres

4
Corner support detail, shaped to funnel rainwater

2

3

4

5

6

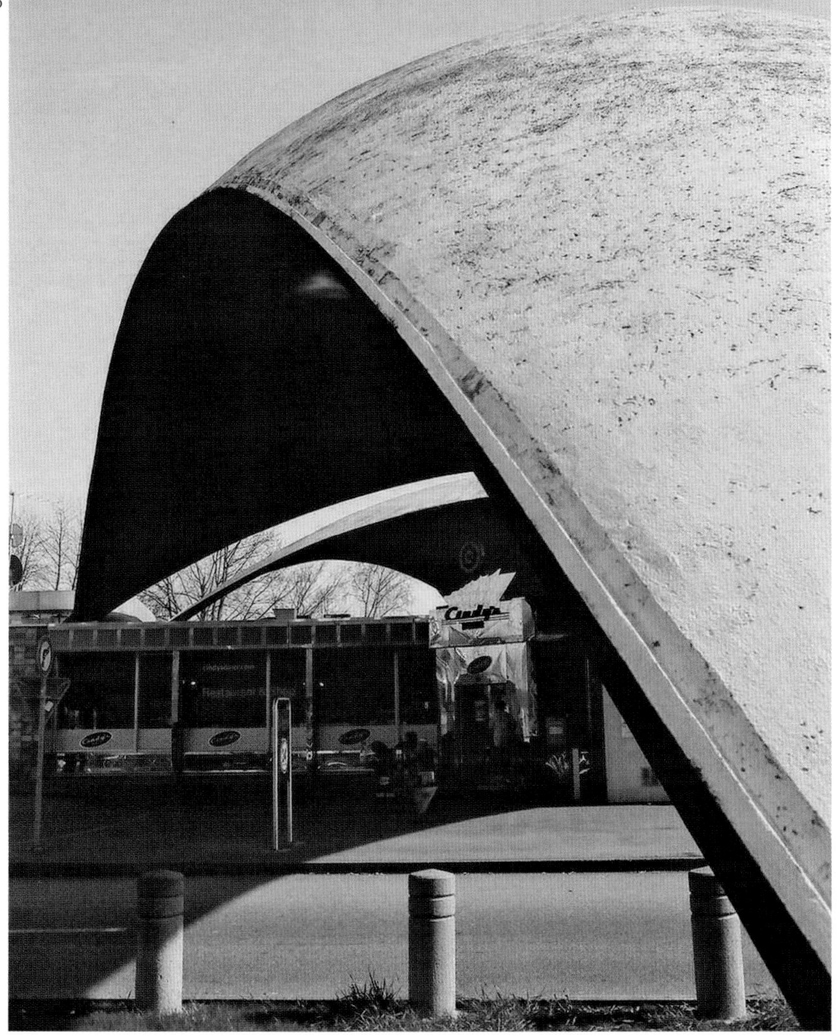

5
Panoramic view of Deitingen Süd Service Station, showing both shells

6
Roadside shell at Deitingen Süd

7

8

7
Solothurn Tennis Centre, showing one bay of the repeated 'hanging membrane' type shell

8
Corner detail of Brühl Sports Centre, Solothurn

9

10

9
Connection detail between two shells at Brühl Sports Centre, Solothurn

10
Interior of Brühl Sports Centre, Solothurn, with central openable rooflight

4.3.7
Jefferson National Expansion Monument ('Gateway Arch')

Structural description
Weighted stressed-skin catenary arch

Location
St. Louis, Missouri, USA

Completion date
1965

Plan dimensions
192m between legs at ground level

Height
192m

Architect
Eero Saarinen (1910–1961) (Eero Saarinen and Associates)

Engineer
Hannskarl Bandel (Severud-Perrone-Sturm-Bandel)

Fabrication and construction
Pittsburgh-Des Moines Steel Company

Although the first Jefferson Memorial design of 1948 was of partly subjective inspiration, it was also a stroke of rational structural functionalism: a catenary arch which, geometrically, was as predictable as a circle.[1]
Allan Temko

Upon his victory in the international design competition in 1947, the winning notification letter was famously sent not to Eero Saarinen but his father, the other 'E' Saarinen: highly acclaimed architect, Eliel. Once this mix-up was resolved, Eero set about assembling a team to design and engineer this monument to the westward expansion of the United States. The structural engineering consultancy of Severud-Perrone-Sturm-Bandel was chosen, and a principal of this office, Dr. Hannskarl Bandel, began to work together with Saarinen to develop the project. Bandel is credited with helping Saarinen achieve the desired geometry of the arch. In a brilliant account of the engineering and construction process, a former colleague of Bandel's, Nils D. Olssen,[2] explains how Saarinen's desire to utilize a hanging-chain 'catenary' curve was transformed when Bandel modelled a catenary curve from differently sized and weighted links, which altered the profile to Saarinen's desired shape. This early modelling seems to have proved extremely useful in developing an equilateral triangular, prismatic cross-section of tapering size, with a flat edge to the back of each leg. At ground level, the legs stand 192 metres apart, which matches the arch's visible (above-ground) height. The arch's external cross-section varies from 16.45 metres at the base to 5.18 metres at the apex. As a structural work, the monument is not only challenging in its geometry but also in its material construction, assembly and structural type. The structure consists of a double skin of steel held together with internal ribs, which forms a type of stressed skin or semi-monocoque structure and dispenses with any separate structural frame – so this is not a steel-clad structural framework; here, the cladding is also the structure. The outer skin was fabricated in 6.3-millimetre cutlery-grade type 304 stainless-steel sheet and the inner skin is 9.5-millimetre carbon-steel sheet. The two arch legs were constructed simultaneously from 142 prefabricated sections. On-site cranes of up to 22 metres in height lifted these segments, after which operation specially designed climbing cranes, mounted to the back of the arch legs, would lift each segment into place. Each section was welded together, with no little skill involved in such long, fully welded seams, which cause local distortion owing to heat. As the construction proceeded, the cantilever of each leg steadily increased, and at 162 metres high a 78-metre stabilizing truss was raised using the climbing cranes and fixed until the arch was complete. The final two 'keystone' segments were designed to be fixed into place very early in the morning when the temperature of the structure was stable. However, when news of this momentous occasion got out, the mayor requested a daylight operation so that it could be recorded for posterity. When sun hit the structure, differential movement in the legs prevented the final connection – a dilemma that was only solved by the attendance of the local fire service, who cooled the back of the arches with sprayed water that caused each leg to slowly rise to the correct position.

Interestingly, the void between the inner and outer steel skins of the arch was filled with concrete up to a height of 91 metres and reinforced with steel tendons; above this level, steel stiffeners were employed. This concrete mass is used to prevent sway and ensure that the thrust line is straight down into the 18-metre-deep foundations, rather than forcing the legs outwards. The concrete also helps to resist buckling, a technique that was utilized in the very slender, rakishly angled columns of Will Alsop's Peckham Library (London, 2000), which were pumped with concrete after they were positioned. In pictures, the Jefferson Memorial is an impressive piece of processed steel. However, what may not be immediately obvious is that this is a visitor 'experience' and, in the best tradition of such edifices – including Eiffel's eponymous tower, the Statue of Liberty (Eiffel-engineered) and London's Great Fire 'Monument' – this is a building for ascending. In what Bandel told Olssen was the real engineering triumph of this project, not-quite-vertical transportational devices take you up to a prismatic interior of seemingly doll's-house proportions, from the windows of which you can view eastwards to the mighty Mississippi River and westwards to St. Louis and beyond. From ground level, you would be hard-pressed to even see these lookout windows. A unique tram system, devised by lift specialist Dick Bowser and comprising five-person pressed-steel capsules on a 'paternoster' type loop, takes you from the underground museum (buried in the slab) to the summit. The arch legs also contain a service lift and emergency-escape stairs.

1 Temko, A., *Eero Saarinen*, New York: George Braziller Inc., 1962, p. 42
2 Olssen, N. D., 'Jefferson National Expansion Memorial (The Saint Louis Arch)' in *Spans* (The Quarterly Newsletter of Inspired Bridge Technologies), third edition, July 2003, pp. 1–3

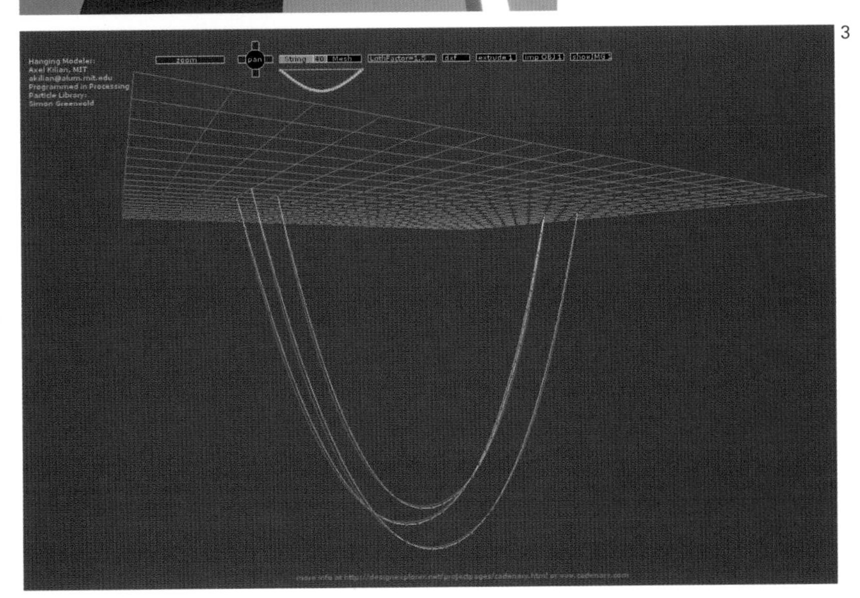

1
St. Louis Arch photographed at night

2
View of the stainless-steel arch from ground level, which shows the tapering triangular cross-section. Note the panel lines, indicating the sheet-steel construction

3
An illustration from CADenary tool v2, a virtual catenary modelling program that has been developed by Dr. Axel Kilian

4.3.8
Maxi/Mini/Midi Systems

Structural description
Steel column-and-truss structures

Location
Switzerland

Completion date
Various (1962–2000)

Plan dimensions
Various

Architect and system designer
Fritz Haller (b. 1924)

The Swiss autodidact architect Fritz Haller has produced three notable steel construction systems, but curiously is still better known for the system furniture he designed for USM. These structural systems, some dating back to the early 1960s, have proved highly effective as flexible and adaptable 'open' systems, and are also quietly structurally innovative.

Haller's three distinct steel building systems are: the Maxi system, for single-storey large-span structures; the Midi, for multistorey, medium-span and densely serviced structures; and his Mini system for one- or two-storey small-span structures. The USM factory in Münsingen utilizes the Maxi system, but the whole facility has been an ongoing project between USM and Haller, which has seen seven phases of construction and expansion between 1962 and 2000. The Maxi system (1963) is the most universal and deliberately open-ended: based upon a 14.4-metre grid, columns are fabricated from four outward-facing rolled-steel-angle sections connected at a distance by steel flats. Large open trusses, also fabricated from standard steel-angle sections sit within the open cruciform column heads to complete the structure. The system is designed to be reconfigurable and easily demountable, and a concise palette of roofing finishes and cladding systems – opaque, glazed, fixed and openable – completes the building envelope. The column configuration is of particular structural interest as lateral stability is cleverly absorbed in moment connections and carefully disaggregated columns, which can incorporate vertical servicing where required, whilst visually the effect is curiously more transparent than might be expected. The column size in the Maxi system is consistent from edge to internal supports (despite different loading conditions) in order to maintain maximum flexibility for future expansion or reconfiguration of these primarily industrial buildings.

Haller's second system was the Mini system (1968), which has been utilized for private residences, small school classes and pavilions. Designed for one- and two-storey structures with spans of 6–7.2 metres, this system uses a mixture of components including steel Square Hollow Sections (SHS) and custom-folded plate-steel elements. Parallels could be made with the work of Jean Prouvé, whom Haller knew, particularly in the use of break-press formed components, which were relatively lightweight in relation to the hot-rolled sections of the Maxi system and were specifically designed for ease of assembly, structural performance and utility. The folded-steel column/mullions (designed to

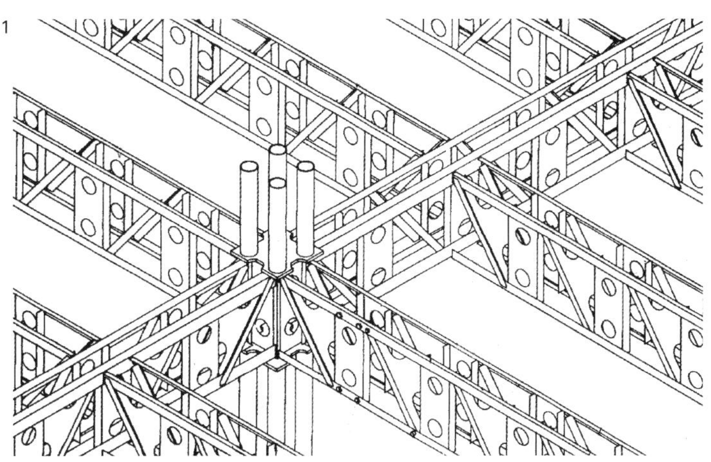

1
Extract from a patent drawing of a variation of the Midi system, 1977

resist shear stresses) work in both the linear condition of a supporting wall and the corner condition, cleverly turning a corner by virtue of their unique profile. Beams are formed from thin folded plate steel and castellated for reduced weight and service runs; the beams also incorporate additional triangular-shaped 'tabs' folded from the flange, to support or fix a soffit or ceiling surface to. The Midi system (1976) is arguably the most sophisticated of Haller's architectural 'products', and combines the use of folded-plate and pressed-metal components and the utility of regular hot-rolled steel sections. Designed with a planning module of 2.4 metres, this is the most open system and can be used for structures of several storeys. Grid configurations of 9.6 x 9.6 metres, 14.4 x 9.6 metres and 7.2 x 7.2 metres, or a mix of these, are possible, with columns relocatable anywhere on that grid. A doubling up of the top and bottom chords and vertical bracing forms a unique truss design. The truss is then stiffened with a specially fabricated folded-and-pressed steel component, which connects all four steel-angle truss chords, thus creating a strong lateral connection that also acts to resist torsional forces. The Midi system has been used for schools, offices and other commercial buildings and represents a higher order of geometric and dimensional coordination, providing for services distribution and maintenance, supports and locations for multiple and easily adapted internal partitioning and simple connections for external envelope and cladding systems. With legislative and regulatory changes in thermal-performance requirements, both the Maxi and the Mini system have thermal-bridge issues that would require design changes. However, the Midi system is still being used for new projects in spite of Haller's retiring from practice, with new schemes coordinated by 2bm architekten.

Interestingly, you will not enjoy any fetishized, large-scale cross-bracing in a Haller project, as lateral stability is cleverly absorbed in moment connections and the carefully disaggregated columns and beams. The lack of visible cross-bracing allows the structural system to remain sufficiently 'open' as to allow major modification, extension or replacement without difficulty owing to lack of structural interdependencies.

2

3

4

2
USM factory: interior of factory showing administrative offices, Münsingen, Switzerland

3
Detail of Maxi system column at building edge, USM factory, Münsingen, Switzerland

4
SBB circular accommodation buildings, utilizing a radial version of the Midi system, Löwenburg, Murten, Switzerland, 1982

5
Temporary school classroom using the Mini system, Solothurn, Switzerland

6
Private residence using the Mini system, 1967, Solothurn, Switzerland

5

6

4.3.9
Tensegrity Structures

Structural description
Tubular aluminium and steel cable tensegrity tower structure

Location
Kröller-Müller Museum, Otterlo, Netherlands

Completion date
1969

Height
30m

Plan dimensions
6m x 6m

Artist
Kenneth Snelson (b. 1927)

The ancient invention of weaving reveals in a direct way the basic and universal properties of natural structure such as modularity, left and right helical symmetry, and elementary structural geometry ... Weaving and tensegrity share the same grounding principle of alternating helical directions; of left to right; of bypasses clockwise and counterclockwise.[1]
Kenneth Snelson

Over the summers of 1948 and 1949, Kenneth Snelson was a student at the unique educational experiment that was Black Mountain College, North Carolina, USA, where staff included composer John Cage, dancer and choreographer Merce Cunningham, painter Willem de Kooning and (most importantly for Snelson) polymath Richard Buckminster Fuller, for whom Snelson began to make models for use in Fuller's lectures. During his time as a student, Snelson developed and formalized the structural innovation of the tensegrity structure, or as Snelson prefers 'continuous tension, discontinuous compression structures',[2] whereby the compression elements of a given structure do not touch each other, insomuch as they are held in space by separate tension elements (strings, wires or cables). There was subsequently much disagreement about the intellectual ownership of this engineering discovery, but both Fuller and Snelson registered patents in relation to tensegrity structures, with Fuller coining the word 'tensegrity', formed from tension and integrity, as one of his composite designed nouns. 'Tensegrity' was included in *The Oxford English Dictionary* in 1985.

The structural interest in tensegrities is more than a vernacular curiosity, as the discontinuity of tensile and compressive forces creates tremendous structural integrity with an even more remarkable material efficiency, most certainly doing more with less and presenting a very useful model of what Snelson calls 'forces made visible'.[3] Within the worlds of architecture and construction, examples of tensegrity structures are thus far relatively limited in number; although the deployment of tensegrity for the Kurilpa Bridge in Brisbane, Australia is impressive, that example may not be the most elegant exemplar of the structural efficiencies integral to tensegrity. Kenneth Snelson has worked as a fine artist since the 1950s and is, through his sculptural commissions and maquettes, the pre-eminent communicator of the potential of the tensegrity structure in all of its forms and configurations and at a number of different scales. Notable works include *Easy Landing* (Baltimore, MD, 1977), which is a horizontal sculpture supported at three points and cantilevered at each end; his *Needle Tower* sculptures I and II (Washington and Otterlo, 1968 and 1971), which are tapering columns made up of 24 progressively smaller (three compressive element) modules; and his *Rainbow Arch* sculpture (private collection, 2001), which creates a semicircular arch using similar three-component modules. In his *Needle Tower II*, Snelson uses a repeated geometric configuration of 24 four-strut tensegrities, but with each module decreasingly scaled. The effect is to make the tower look even taller than its considerable height of 30 metres. The modules at the top of the tower more closely resemble Snelson's smaller-scale structural sculptures, whereas the base module uses building-construction sized elements, none of which appear to have suffered any kind of weathering or fatigue since their installation over 40 years ago.

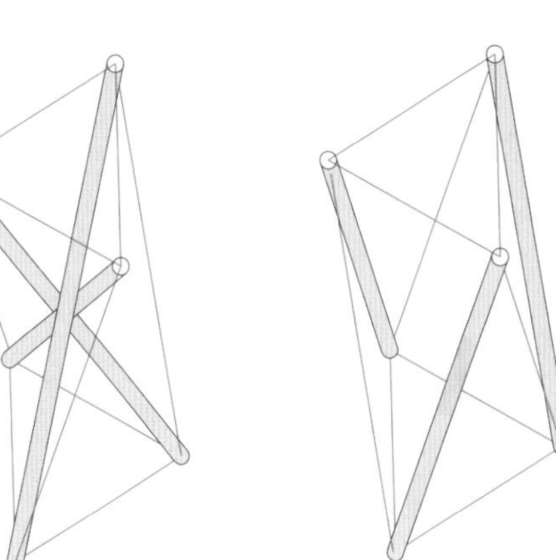

1
Needle Tower II, Kröller-Müller Museum, 1969

2
Needle Tower II during annual cleaning, 2011

3
Two configurations of a simple three-strut tensegrity structure, where the compressive struts (and thus the forces) are both connected and held apart with tensile wires

If Snelson has extended structural possibilities through sculpture, then Fuller's thought experiment about the potential for such structures is equally inspiring. Musing on the structural qualities of the rim-and-spoke bicycle wheel, it seemed to Fuller that this was perhaps the most ubiquitous instance of 'tensional integrity where tension was primary and comprehensive and compression secondary and local'.[4] Fuller saw the possibility of applying the tensegrity principle at various scales, and posits the idea of replacing the wheel's compressive struts, or members, with miniature tensegrity structures, and the struts within the miniature tensegrity masts replaced by even smaller tensegrity masts, and so on until you reach molecular-sized manipulations. 'At this stage of local miniaturization the inherent discontinuous-compression, tensional integrity of the non-solid atomic structures themselves would coincide with the overall structuring principle of the whole series of masts-within-masts complex, thus eliminating any further requirements of the now utterly obsolete conception of "solid" anything'.[5] The cell biologist and founding director of the Wyss Institute, Don E. Ingber, has made the connection between the tensegrity structures of Snelson and living cells, and asserts: 'An astoundingly wide variety of natural systems, including carbon atoms, water molecules, proteins, viruses, cells, tissues and even humans and other living creatures are constructed using a common form of architecture known as tensegrity.'[6] Ingber summarizes the operational characteristics of tensegrities thus: 'Tensegrity structures are mechanically stable not because of the strength of individual members, but because of the way the entire structure distributes and balances mechanical stresses.'[7] And so, although this structural principle is a rarely deployed commodity in the construction industry, its inherent strength and potential lightness offer huge possibilities in the fields of structural engineering, architecture and beyond.

1 http://www.kennethsnelson.net/icons/struc.htm (accessed 20.9.12)
2 Heartney, E., *Kenneth Snelson: Forces Made Visible*, Lennox, MA: Hard Press Editions, 2009, p. 22
3 Op. cit., p. 9
4,5 Krausse J., and Lichtenstein C., *Your Private Sky: R. Buckminster Fuller*, Zürich, Lars Müller Publishers, 2001, p. 232
6,7 Ingber, D. E., 'The Architecture of Life' in *Scientific American*, January 1998, pp. 48–57

4.3.10
Munich Olympic Stadium Roof

Structural description
Mast-supported cable net

Location
Munich, Germany

Completion date
1972

Roof area
34,500m^2

Height of tallest mast
80m

Architect
Günter Behnisch (1922–2010) with Frei Otto (b. 1925)

Engineers
Fritz Leonhardt, Jörg Schlaich and Heinz Isler

Frei Otto not only considers the temporary nature of his membrane structures desirable, but admits that his objections to making architecture stem from his reluctance to fill the earth's surface with lasting buildings. He hesitates to pursue a project unless he is certain that its realization will be temporary enough not to be in man's way. This endorsement of obsolescence contradicts the traditional view of architecture as a fulfilment of man's need for monuments. Yet, as vernacular buildings of all periods prove, artistic value is not dependent on the durability of a structure, nor on the amount of preciousness of its material. On the other hand, temporariness does not mean improvisation, as is evident from the amount of research invested in each lightweight structure.[1]
Ludwig Gläser

Given the above, it seems contradictory that this most celebrated work of Frei Otto no longer belongs to the category of temporary or ephemeral structures, having been designated as national protected monument in 2000. It may also be worth noting that Otto was not even involved in Günter Behnisch's winning competition entry of 1967, although its design and technology were clearly influenced by Rolf Gutbrod and Otto's recently completed German Pavilion at the Montreal Expo in April 1967. When the technical feasibility of the competition winner was subsequently called into question, Frei Otto was contacted by Behnisch, and, working with his Institute of Lightweight Structures (IL) in Stuttgart, Otto developed the final form for the stadium roof.

This colossal roof structure consists of nine interconnected 'anticlastic' (or saddle-shaped), curved cable nets, which are supported by welded tubular-steel masts up to 80 metres long and with a 50-meganewton load capacity. The masts, which puncture the roof membrane, are positioned behind the spectators at the rear of the west stand, and they support, or 'pick up' the skin of the roof at two points with suspended cables. The front edge of the roof is held taut by a continuous edge cable, pulled across the structure and anchored to the north and south of the stadium. The technical challenges of an innovative project like this were numerous – not least coping with the massive tensile forces required to act on the cable net, keeping it in place. The two biggest tensile loads at the front edge, with pulls of up to 50-meganewtons, were resisted by inclined-slot and gravity-anchor foundations, which formed massive buried concrete diaphragm walls using opposing geometry and mass to resist the tensile forces. Elsewhere in the stadium, ground anchors were used to resist tensile forces, a technology untried in Germany at that time. The cable-net surfaces themselves were formed by a rectangular grid of paired cables, of either 11.7 or 16.5 millimetres diameter. The grid dimension was 750 millimetres; however, Otto was not happy with this,

1

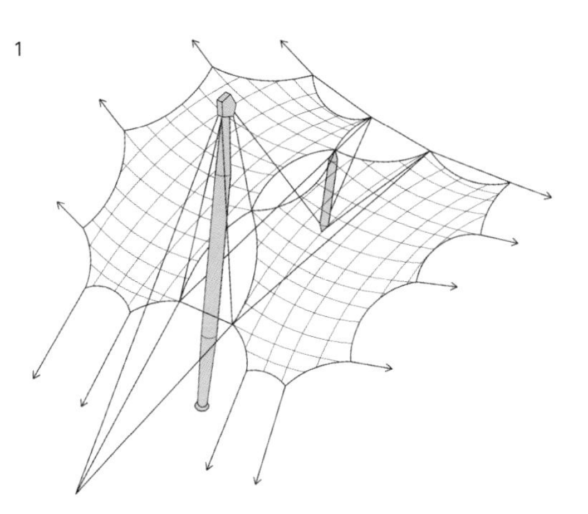

1
Diagram of two bays of the Munich Olympic stadium roof showing how the anticlastic roof surface is created with a mast-supported cable net pulled down to ground at the back with the free edge supported by a longitudinal tensile cable

arguing that a 500-millimetre grid would be considerably safer during construction. The cables are fixed together at intersections with aluminium clamps, which allow them to rotate in relation to each other when pulled into the final configuration. The edge cables and main support lines are all in 80-millimetre-diameter steel cable, with the front edge consisting of a bundle of these elements clamped together in cast-steel 'arms'. The cable net was eventually clad in 3 metre x 3 metre x 4 millimetre-thick clear acrylic panels fixed by flexible neoprene connectors at the cable-intersection nodes, with the joints between the panels sealed by a neoprene strip clamped to the panel edges. The weathering strips are, curiously, one of the most visible delineations of the structural form, although they are non-structural. The original design had investigated cladding the cable net in a PVC membrane, timber sheathing or even thin pre-cast concrete panels. Otto has subsequently constructed cable-net structures that are entirely clad in glass.

Frei Otto had first developed cable-net structures in the early 1960s, when his work with fabric structures began to become dimensionally limited by the tensile strength of a given substrate. By disaggregating the tensile forces into a low-resolution weave of fewer but stronger fibres (typically steel cabling), Otto could achieve considerably larger structures, which were first seriously prototyped at the Montreal Expo in 1967. The cable net forms a structural grid, which is then clad – in the case of Montreal, largely in fabric. Cable nets are certainly not the only structural innovation of Frei Otto, who pioneered the use of tensile fabric structures and developed a formidable array of pneumatic and branching structures. What is particularly impressive about his work are the form-finding techniques he developed to model these hitherto unimaginable structures. In particular, Otto developed soap-bubble modelling, wherein the fine meniscus of a soap film finds its form within a geometrically delineated frame. This type of prototyping was born out of Otto's close observation of nature and natural processes in a way that pre-dates the development of biomimetic engineering, whereby engineers define solutions through the study of natural processes (human, animal and organic).

Otto's experimental work, carried out with students of the IL in Stuttgart, are particularly well documented in the *IL Documents*, a series of books published between 1969 and 1995, which investigate specific material, structural and geometric properties. This substantial body of research is unique in that the ambitions of the work are neither exclusively engineering nor design, but a synthesis of the two.

The Munich Olympic Stadium remains a remarkable achievement, which must have seemed startling 40 years ago. Frei Otto remains one of the very few figures whose interest in structural innovation and experimentation outweighs his ambitions as a builder. In a lecture at the Architectural Association in the late 1990s, Otto explained to a questioner that, owing to the nature of his constructions – which might be a tent or an inflatable – he was never entirely sure of the location or number of Frei Otto buildings in existence on the planet at any given moment.

1 Gläser, L., *The Work of Frei Otto*, New York: MoMA, 1972, p. 10

4.3.10
Munich Olympic Stadium
Roof

2

3

2
Munich Olympic Stadium: view of main stand

3
Detail of tubular-steel compression mast at the rear of the stadium

4
Roof covering to the rear of the main stadium

5
Detail of cable-net and polycarbonate panel connection

6
The cable-supported roof incorporates floodlight rigs. Tours of the stadium include a walk on the roof edge

7
The tensile roof-edge element comprises a cluster of ten separate woven-steel cables

4

5

6

7

4.3.11
Bini Domes – inflatable formwork

Structural description
Reinforced-concrete dome, utilizing inflatable formwork

Location
Killarney Heights, New South Wales, Australia

Completion date
1973

Height
5.5m

Plan dimensions
18m diameter

System designer
Dr. Dante Bini (b. 1932)

Architects
NSW Department of Public Works with Dr. Dante Bini

Engineers
Taylor, Thompson and Whitting Consulting Engineers with Dr. Dante Bini

For over 45 years, Italian-born architect Dr. Dante N. Bini has dedicated his professional life to the development of what he calls 'automated construction technologies'. In 1965, in Bologna, Italy, he successfully constructed a 12-metre-diameter, 6-metre-high hemispherical concrete shell structure in three hours, using the unique pneumatic formwork of a giant balloon. This first prototype did, however, have some teething problems, particularly the uneven distribution of the wet concrete caused by an unpredictable (asymmetric) inflation. Improvements were made, and in 1967 at Columbia University, New York, Bini demonstrated in two hours the construction of another large-scale 'Binishell'. For this first US prototype, Bini utilized a complex web of helical 'springs' with steel reinforcement bars threaded through their middle, which allowed for a geometrically controlled inflation and thus a uniform concrete distribution across the shell structure. For this demonstration and subsequent Binishell structures, an additional external membrane was also used, which allowed for the subsequent vibration and compaction of the concrete, post-inflation. Over 1,500 Binishells were constructed throughout the world between 1970 and 1990, with diameters of between 12 and 36 metres and with a varying elliptical section.

Less interested in the experimental form finding of Swiss engineer Heinz Isler's elegant European shells, Bini was concerned with how the construction process itself could evolve and how a lightweight and low-cost resource such as air could be utilized in the construction industry. Concrete shell structures like Isler's and Félix Candela's are structurally efficient and enclose huge volumes with a small amount of material, but the fabrication of formwork required a large on-site semi-skilled workforce. Bini's inflatable formwork, or 'Pneumoform', eradicates the need for such a large site team and allows for more high-speed construction.

The sequence of fabrication first involves the construction of a ring beam and ground-floor slab. The ring beam cleverly contains a 'cast in' egg-shaped void, which will contain a separate inflatable tube to hold the main membrane in place during inflation as well as air inlets and outlets. The internal 'pneumoform' of nylon-reinforced neoprene is then laid over the slab and secured at the edge; on top of it, a complex network of criss-crossing helical springs is stretched across the diameter of the circular ground slab. The springs have no specific structural function but control the even distribution of steel reinforcement bars, which are threaded through the springs, and also maintain an even concrete thickness by holding the mix in place. Once the reinforcement is in place, the concrete is poured. A regular concrete mix is used with small amounts of retarders and plasticizers added to extend the workability of the mix for two to three hours. After the pour, an outer membrane of PVC is laid over the wet concrete, which will help to control evaporation during the setting process and allow for vibration of the concrete. The inflation procedure then begins, using low-pressure blowers, and takes about one hour; pressure is regulated by controlling the outlet to maintain an even 'lift'. When the shell is fully inflated, the concrete is vibrated using rolling carts hung from cables at the top of the structure. The internal air pressure is maintained for between one and three days depending on the diameter. For a 36-metre-diameter dome, the thickness of the completed shell is 125 millimetres at the base and 75 millimetres at the crown.

Critical to the success of this innovative construction technique and structural type was the system design and fast construction programme. Bini designed the 18-metre-diameter dome for Killarney Heights Public School, New South Wales, to be erected (with foundations already in place) in 12 days. On the tenth day the concrete-covered membrane was inflated and subsequently vibrated free of air pockets with the innovative guided vehicles (described above). By day 12, the reinforced concrete shell was sufficiently stable to begin to cut openings for entrances, windows and ventilation.

1–4
Construction of Killarney Heights Public School Binishell, New South Wales, Australia, 1973

5
Completed building

1

2

3

4

5

4.3.12
Niterói Contemporary Art Museum

Structural description
Cylindrical cantilever

Location
Niterói, Rio de Janeiro, Brazil

Plan dimensions
50m diameter at roof level

Architect
Oscar Niemeyer (1907–2012)

Engineer
Bruno Contarini

Completion date
1996

Height
16m

1

Oscar Niemeyer was in his eighties when he designed the Niterói Contemporary Art Museum along with his long-time collaborating engineer, Bruno Contarini.

The building consists of three floors built into a cupola that cantilevers from a cylindrical base. The base springs from a reflecting pool, and the cupola is accessed by a snaking ramp. The building is constructed from reinforced concrete and employs three circular floorplates ranging from 36 to 40 metres in diameter and supported by a central cylinder 9 metres in diameter. The floorplates employ prestressed girders resting on 50-centimetre-diameter columns.

Each sheet of the seventy 18-millimetre-thick triplex glass plates is 4.80 metres high and 1.85 metres wide. Framed by steel bars and with an inclination of 40 degrees to the horizontal plane, they can sustain a load equivalent to 20 people.

The structure was designed to withstand a weight equivalent of 400 kilograms per square metre, and winds of up to 200 kilometres per hour. It consumed 3.2 million cubic metres of concrete.

2

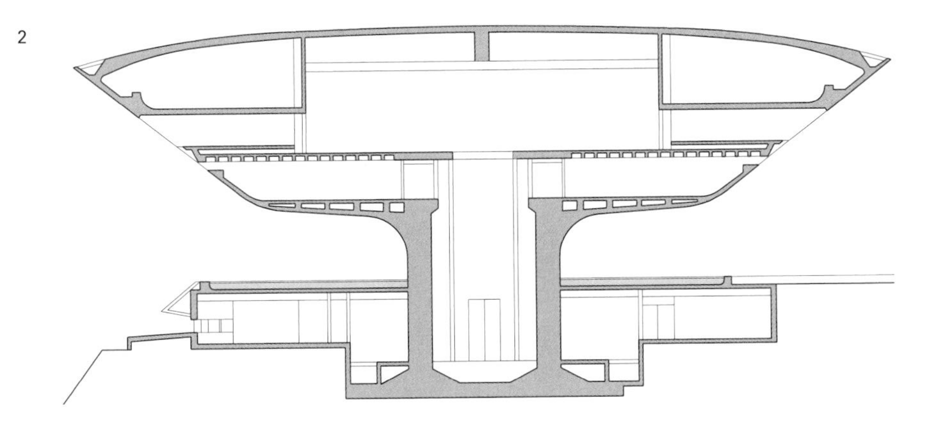

3

4

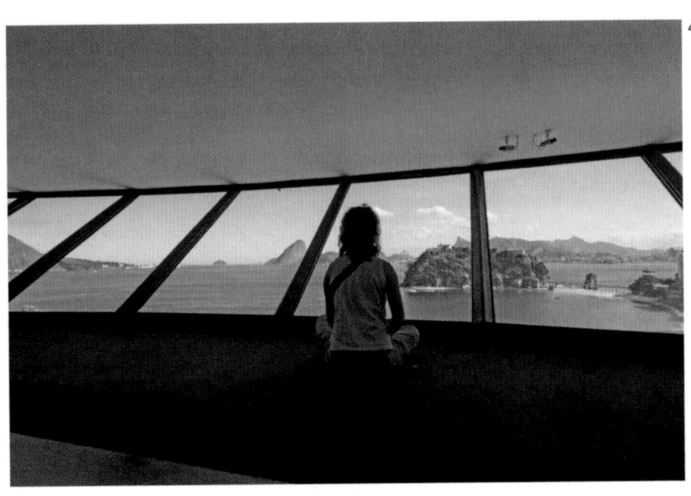

1
Niterói Contemporary Art Museum

2
Cross-section through the museum

3
Detail of the central cylinder base and reflecting pool

4
Interior view looking out to Guanabara Bay

4.3.13
Structural Glass

Structural description
Load-bearing glass structures

Locations
Various

Completion dates
1990–1997

Engineer
Tim Macfarlane (b. 1954)

Glass is no longer an ornamental item ... but has emerged into a structural element.[1]
Fazlur Khan

In the early 1990s, there was a quiet revolution in the way that glass was employed in architecture as a structural material. This increased experimentation in the application of glass was not limited to the thin sheath of the building skin – framed in timber, steel or aluminium – but increasingly extended to frameless glazing and, ultimately, to structural glazing with no support at all other than crafted laminations of glass itself and the magic of structural silicone. At the forefront of these new approaches to the art, architecture and specifically the engineering of these experimental and innovative projects was the structural engineer Tim Macfarlane of Dewhurst Macfarlane Consulting Engineers.

Through a series of small but iconic projects in close collaboration with architects such as Rick Mather, Eva Jiricna and Ohlhausen DuBois Architects, Macfarlane helped to change the way in which glass was classified as a construction material and redefined the engineering potential of this wondrous substrate. He likens this process to 'making rules up as you go along' insomuch as the structural properties and material-performance expectations were not comprehensively codified as part of the structural investigations. Macfarlane also draws parallels with the proliferation and wonderful diversity of reinforced-concrete use as architects and engineers began to test the limits of this new material at the beginning of the twentieth century. From Maillart, to Luigi Nervi, to Félix Candela (to name but three), these 'structural artists' were not reading rule books but writing them, each in his own highly individualized way and for differing programmatic instances. After this flowering of diverse and intriguing engineering approaches, Macfarlane suggests that a kind of Fordism took over and industrial efficiency tended to normalize and limit possibilities. With industry less likely to be 'light on its feet' and more likely to play an increasingly protectionist game, the possibilities were limited through a codification of structural properties linked to relative economic success and known methods of construction.

The reliance on a mathematical model to create a design is only one approach, and Macfarlane states: 'Maths has never led me to a solution, but has helped to determine how to represent the solution'.[2] Macfarlane also adds that the full extent or knowledge of a material and its properties are virtually unfathomable, and therefore structural possibilities and strategies should not be limited by our own experience.

Macfarlane categorizes a brief history of his

1

1
The Klein Residence, Santa Fe, New Mexico, by Ohlhausen DuBois Architects, uses glass as a primary load-bearing element (for description see overleaf)

own work, and the technological development of structural glass, with a series of projects, prototypes and material tests that are detailed overleaf. These range from simple, lateral innovations in fabrication or assembly to completely new methods of construction using glass. Macfarlane cites the advent of the consulting engineer, formalized between 1907 and 1915, as an important evolutionary stage in the proliferation of structural possibilities. These possibilities are, by definition, only limited by our knowledge of material properties, fabrication and assembly techniques, as well as other instruments of structural advantage such as geometry. However, Macfarlane thinks that it is only through the full exploration of these fields that architects and engineers can challenge the intellectual-property-protected 'products' of patented systems of construction and better answer the detailed programmatic requirements of any given job with hitherto unimagined structural and engineering solutions.

1 Khan, Y. S., *Engineering Architecture: The Visions of Fazlur R. Khan*, New York: Norton, 2004, p. 79

2 Interview with Tim Macfarlane by Will McLean, 3 May 2012

Joseph shop

Structural description
Tensile steel rods and structural glass frame

Location
London, England

Completion date
1990

Architect
Eva Jiricna (b. 1939)

Engineers
Dewhurst Macfarlane

A seemingly small but important innovation allowed this intricate and elegant staircase to have its trademark transparent stair treads. In each case, Macfarlane layered together, but did not laminate, a 19-millimetre sheet of sandblasted annealed glass and a 15-millimetre-thick piece of acrylic. The glass provided stiffness and a hardwearing top, the acrylic a safety factor.

2
Joseph shop staircase with layered glass stair treads and stainless-steel rods

Klein Residence

Structural description
Glass as primary load-bearing element

Location
Santa Fe, New Mexico

Completion date
2007

Architects
Ohlhausen DuBois Architects

Engineers
Dewhurst Macfarlane

The Klein Residence represents an audacious approach to structural glass. Its use for the house's glass lookout pavilion is both 'double-take' inducing and a thoroughly worked engineering solution. The aim was to create a living room with uninterrupted views towards the mountains without any visible structural impediment. The result is a space where the two glazed sides of the living room meet in the northwest corner with the steel structure of the roof supported by the glass alone. The architect Mark DuBois has stated that 'the architectural space formed by the load-bearing glass wall is visually remarkable and psychologically very intriguing'.[1] After initially exploring the option of an all-glass corner column (L-shaped or cruciform in plan), the design team proceeded with the concept of an all-glass multi-panel bearing wall. The wall, 3.5m high by 8.6m long, comprises seven equally sized panels and makes up the west wall of the room. The adjacent north wall (also fully glazed) is visually identical, but non-loadbearing. Each structural glass panel is fabricated from three sheets of fully tempered (toughened) glass laminated with PVB film. The central sheet is 19mm thick, with 6mm sheets each side. The two outside sheets are slightly shorter so that all load travels through the central sheet. The structural glass wall was engineered with a safety factor of three and designed to have a maximum deflection of L/100, which is 35mm over the 3.5m height. To avoid any visible framing at the head and sill of the glass, a special steel channel was fabricated and recessed into the floor and soffit. The success of the engineering concept depends on an even distribution of the static load throughout the seven panels; this became one of the main challenges for the design and engineering team. The solution was to make the steel channel adjustable, using threaded rod at the top and bottom of the glass panels. Stacks of spring washers were used at the roof connection to further ensure equitable support along the length of the wall and redistribute that load in the case of a panel failure. If previous developments of structural glass have produced remarkable 'all glass' structures, then the Klein Residence shows how glass can be utilized as a structural support system for other (non-glass) elements.

1 http://www.boishaus.com/glass_performance_days_2007.pdf (accessed 20.9.12)

All-glass Extension

Structural description
Laminated all-glass beam and column structure

Location
London, England

Completion date
1992

Architect
Rick Mather

Structural engineers
Dewhurst Macfarlane

This extension to a private residence, although relatively modest in scale, has had an enormous impact on the perception and expectations of glass technology in architecture. This is a lean-to structure, in which the columns and beams comprise laminations of three 12-millimetre-thick sheets of annealed glass bonded together with clear resin. The beams are cut to a curved profile, and are 275 millimetres deep at their midpoints and 200 millimetres deep at the column connection, which is a mortise-and-tenon joint (see Broadfield House, below). The columns, which are 200 millimetres deep, are similar laminations to the beams, and this layering provides an inbuilt safety factor. The structure is clad in double-glazed units that uniquely feature glass-edge spacers for increased transparency, and the roof panels are coated with a conductive layer that can be used as a heating element.

3
All-glass extension showing laminated glass beams and columns

Broadfield House Glass Museum

Structural description
Laminated all-glass beam and column structure with all-glass box beam

Location
Dudley, England

Completion date
1994

Architect
Design Antenna

Engineers
Dewhurst Macfarlane

This all-glass structure was built as an extension to Broadfield House Glass Museum in Dudley. The glass technology is similar to that used in Macfarlane's earlier All-glass Extension with Rick Mather, but this project is significantly larger, with the structure measuring 11 metres long x 5.7 metres wide x 3.5 metres high. The glass columns and beams are 32 millimetres thick, and made from three layers of 10-millimetre annealed glass bonded with a resin laminate. The beams and columns are connected at the top edge by a mortise-and-tenon joint, with the centre layer of three laminations protruding from the column and the central layer of the beam cut back accordingly. The columns are at 1,100-millimetre centres and are 200 millimetres deep, with the beams 300 millimetres deep. The double-glazed cladding panels of the front face and roof are bonded to columns and beams with structural silicone. Another intriguing feature of this project is the 2.2-metre-wide opening created for glass doors, which is achieved using an all-glass box beam or lintel – surely another structural first.

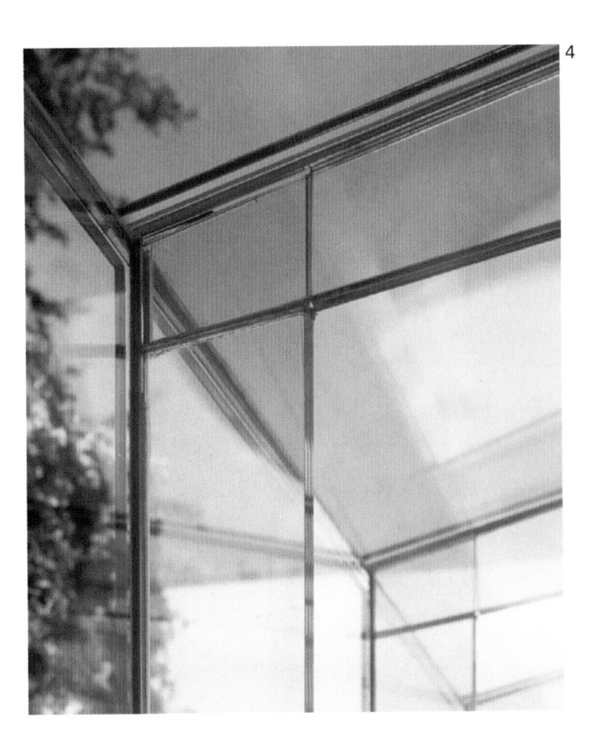

4
All-glass extension showing mortise-and-tenon joint between column and beam

Station Entrance Canopy

Structural description
Cantilevered glass beams in four offset sections

Location
Yurakucho, Tokyo, Japan

Completion date
1997

Designers and engineers
Dewhurst Macfarlane

On the plaza of Rafael Viñoly's Tokyo International Forum, Tim Macfarlane was invited to submit a design for a canopy to Yurakucho underground station. What he designed, engineered and ultimately built was an unprecedented 10.6-metre-long x 4.8-metre-wide cantilevered canopy, fabricated entirely from glass. The glass roof is supported by three glass composite beams, which each consist of four groupings of glass blades that taper from the cantilever connection to the unsupported edge. The glass-beam components consist of two 19-millimetre-thick glass sheets, laminated together, which are bolted at the midpoint and at the end of the next offset group of glass 'blades'. The number of laminated, layered glass components is four at the steel cantilever connection and reduces to a single glass component (of two laminated layers) at the canopy top edge.

The mechanical connections between the components are made with 50-millimetre-diameter high-strength stainless-steel pins, with specially designed bezels fitted to the holes for a more even load distribution. What made this project technically feasible was a combination of the physical testing carried out with glass fabricators Firman Glass and City University, and Finite Element Analysis. Although the results of this glass testing had been successful, the clients decided to also use acrylic as beam components as an additional safety factor; these elements are only visible through their different edge colour, which is much lighter than that of glass. The outer canopy skin is made from a lamination of two 19-millimetre glass sheets, with joints bonded and sealed with structural silicone.

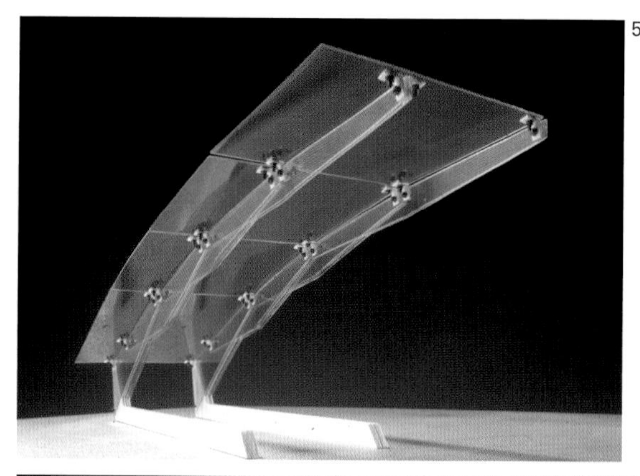

5
Yurakucho Station canopy model

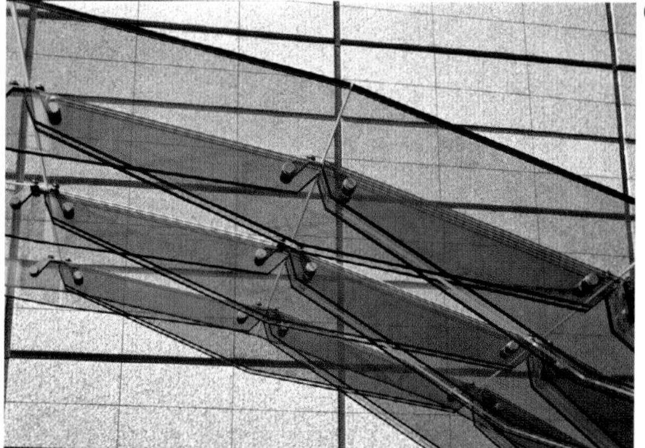

6
Detail of Yurakucho Station canopy showing overlapping glass beam connections

Apple Stores

Structural description
Laminated glass panels and all-glass reciprocal beam system

Locations
Various

Completion date
2006

Architects
Bohlin Cywinski Jackson

Designers and engineers
Dewhurst Macfarlane

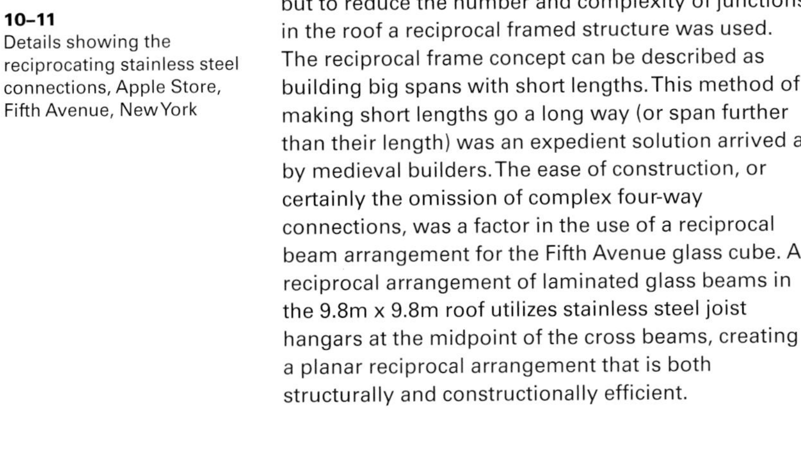

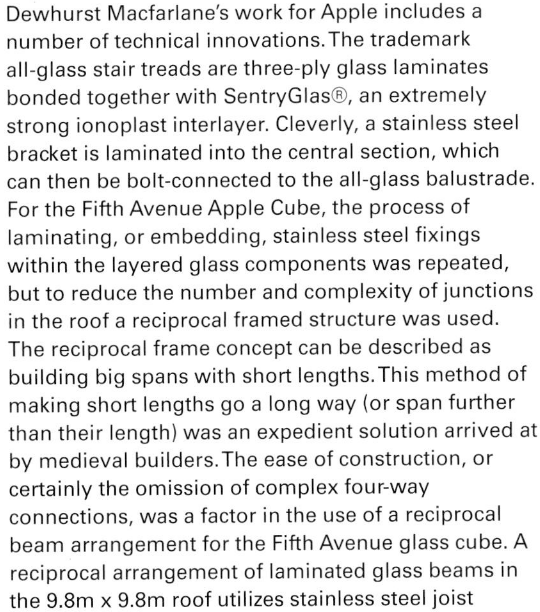

Dewhurst Macfarlane's work for Apple includes a number of technical innovations. The trademark all-glass stair treads are three-ply glass laminates bonded together with SentryGlas®, an extremely strong ionoplast interlayer. Cleverly, a stainless steel bracket is laminated into the central section, which can then be bolt-connected to the all-glass balustrade. For the Fifth Avenue Apple Cube, the process of laminating, or embedding, stainless steel fixings within the layered glass components was repeated, but to reduce the number and complexity of junctions in the roof a reciprocal framed structure was used. The reciprocal frame concept can be described as building big spans with short lengths. This method of making short lengths go a long way (or span further than their length) was an expedient solution arrived at by medieval builders. The ease of construction, or certainly the omission of complex four-way connections, was a factor in the use of a reciprocal beam arrangement for the Fifth Avenue glass cube. A reciprocal arrangement of laminated glass beams in the 9.8m x 9.8m roof utilizes stainless steel joist hangars at the midpoint of the cross beams, creating a planar reciprocal arrangement that is both structurally and constructionally efficient.

7
Apple Store all-glass stair, Chicago, 2010

8
All-glass stair detail showing bolted stair treads, Apple Store, Chicago, 2010

9
Glass cube, Apple Store, Fifth Avenue, New York, 2006

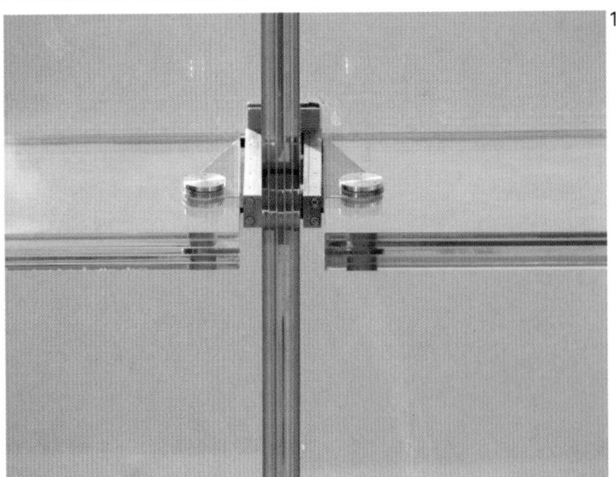

10–11
Details showing the reciprocating stainless steel connections, Apple Store, Fifth Avenue, New York

4.4
2000–2010

4.4.1
Ontario College of Art and Design expansion, featuring the Sharp Centre for Design

Structural description
Steel-truss box

Location
Toronto, Ontario, Canada

Completion date
2004

Plan dimensions
Steel-truss 'box' 85m long x 30m wide x 10m high

Floor area
8,400m^2

Height of tapered columns
26m

Architects
William Alsop (b. 1947) with Young + Wright Architects

Engineers
Carruthers & Wallace Ltd

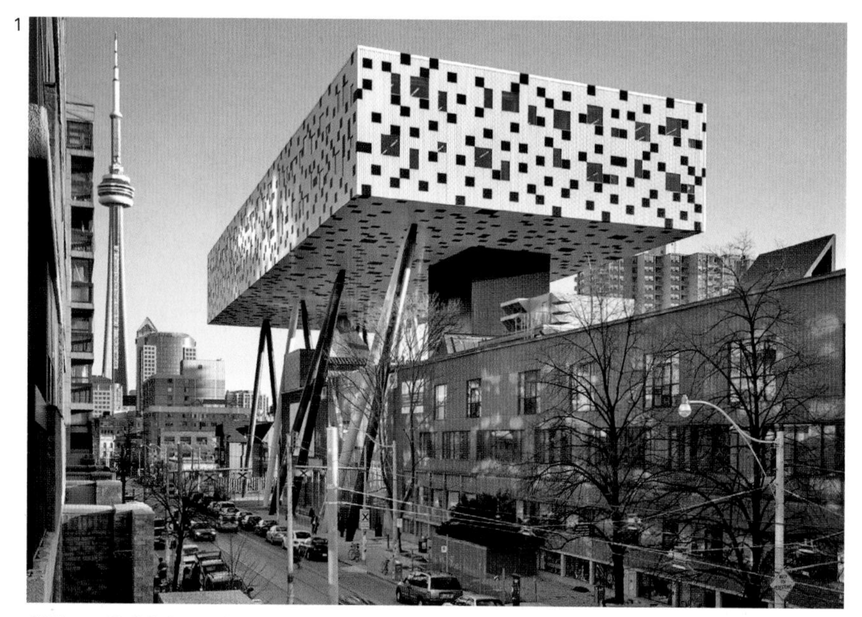

1
Ontario College of Art and Design (OCAD), Sharp Centre for Visual Art: view looking south towards the CN Tower

2
View looking north

When British architect William Alsop was invited to design an extension for the Ontario College of Art and Design (OCAD), he spurned the adjacent site set aside for the scheme and instead elevated this new department in a pixelated aluminium-clad steel box eight storeys above the existing structures, propped on pencil-thin canted legs.

The starting point for the structural engineers was to create a 'tabletop': a stiff, inhabitable structure supported on legs. The stiffness is afforded by two-level structural-steel trusses that span in an east–west direction and are shaped by large diagonal members running through them, which are designed to allow for the passage of both people and services. Between these large two-storey-height assemblies run longitudinal structures that link the horizontal trusses together and provide the perimeter box of the 'table'. To make this box sufficiently stiff, the structure is braced horizontally at the levels of the main and intermediate floors and the roof. The engineers worked closely with the architects in positioning and orientating the 30-metre-long legs in a seemingly random arrangement that also makes structural sense. The triangle is a famously efficient and, more importantly, structurally stable shape, and the leg pairs were thus designed as a series of triangular supports. Another consideration was creating stable elements at ground level to support these legs. The leg-support structures, which extend from the ground level down into the underlying rock, comprise concrete caissons (piles) that range from 900 millimetres to 2 metres in diameter, and extend a distance of up to 18 metres into the rock. The caissons are configured in a triangular pattern (for each pair of columns) and interconnected at grade level to form a three-dimensional frame. Another very important structural consideration is the design of the tabletop to resist lateral loads, which come from two sources in downtown Toronto: wind loads and earthquakes. These lateral loadings are

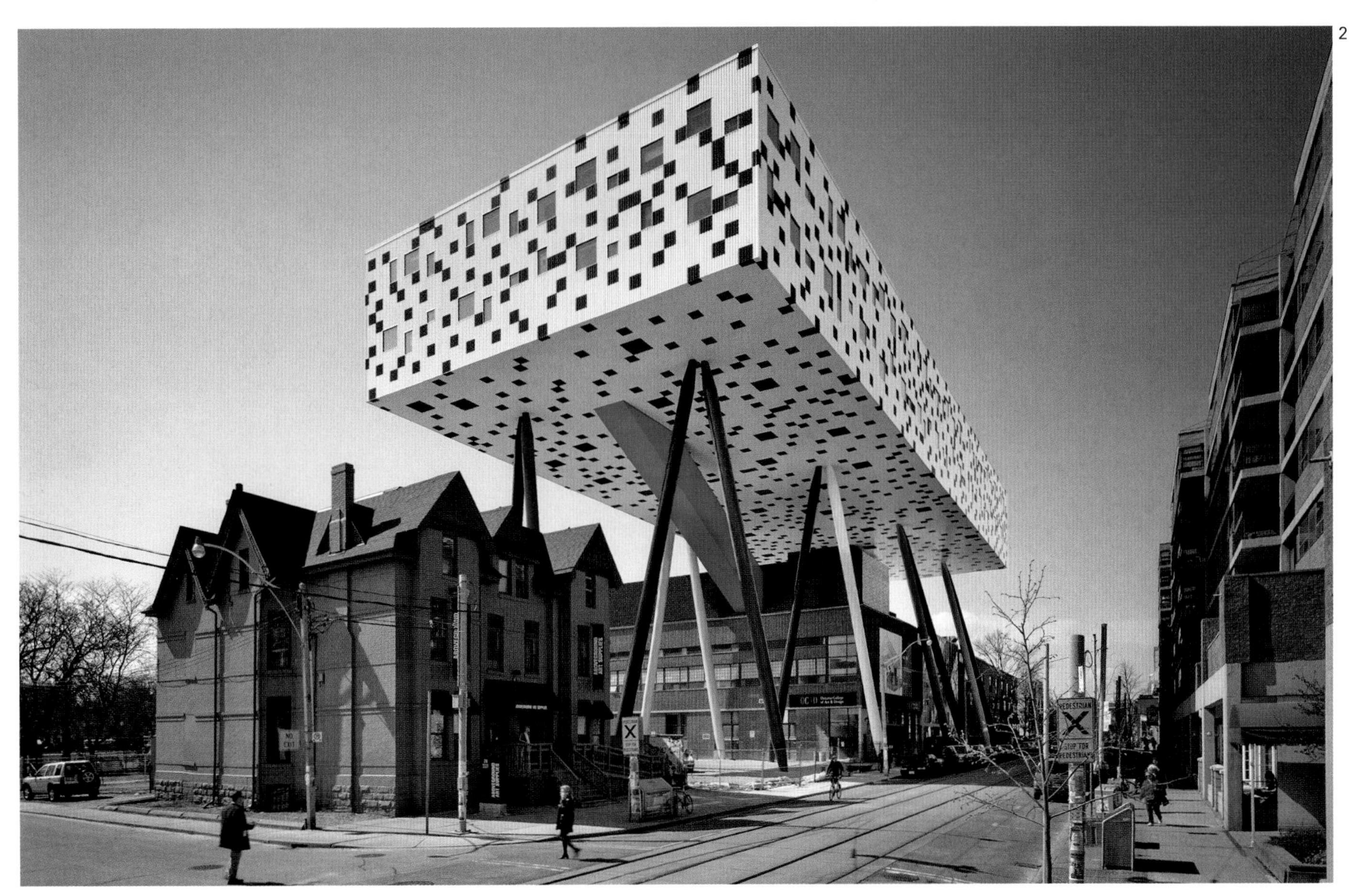
2

resisted in two ways: one is the orientation of the triangular leg elements, which are most effective at resisting lateral loads in the transverse direction; the other is the large, stiff, cantilevered, concrete stair-core element positioned at the northern end of the building, which resists most of the lateral loads in the longitudinal direction.

The tabletop is supported by six pairs of legs; the architect wanted what he called 'cigar legs', which were created by using a large steel Circular Hollow Section (CHS) with specially rolled fabricated-steel conical components welded to each end. These leg elements worked well structurally but were large and heavy items, which carried logistical implications. To avoid unnecessary transportation costs and complex site operations, the steelwork was designed and fabricated in pieces that could be trial-pre-assembled in a workshop and then subsequently reassembled on site. The hollow structural-steel 'cigar legs' are 27 metres in length and 914 millimetres in diameter, with a wall thickness of 25 millimetres. The computer structural model used to evaluate static and live loads estimated a maximum horizontal displacement of 8 millimetres at the southeastern corner of the 'tabletop'. The structural design also includes redundancy, to provide alternative load paths in the event of the catastrophic failure of a leg support.

4.4.1
Ontario College of Art and Design expansion, featuring the Sharp Centre for Design

3

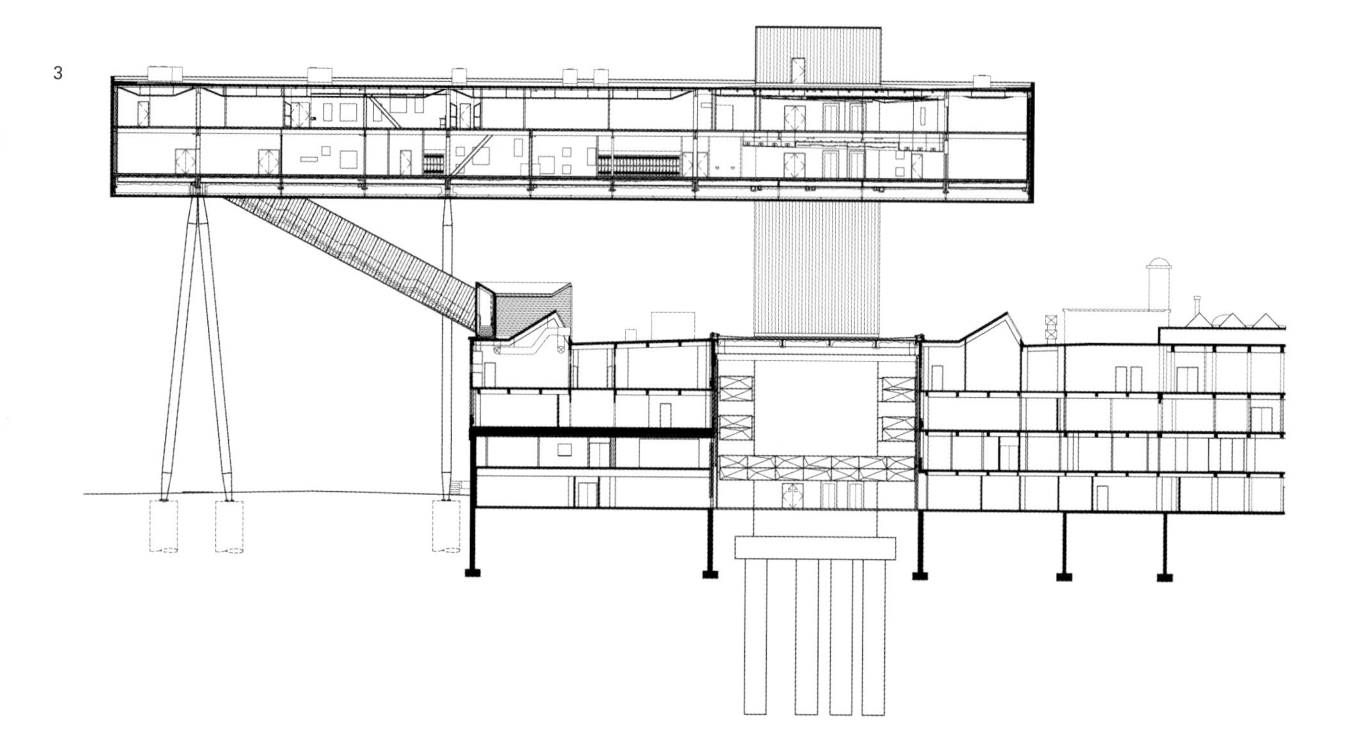

4

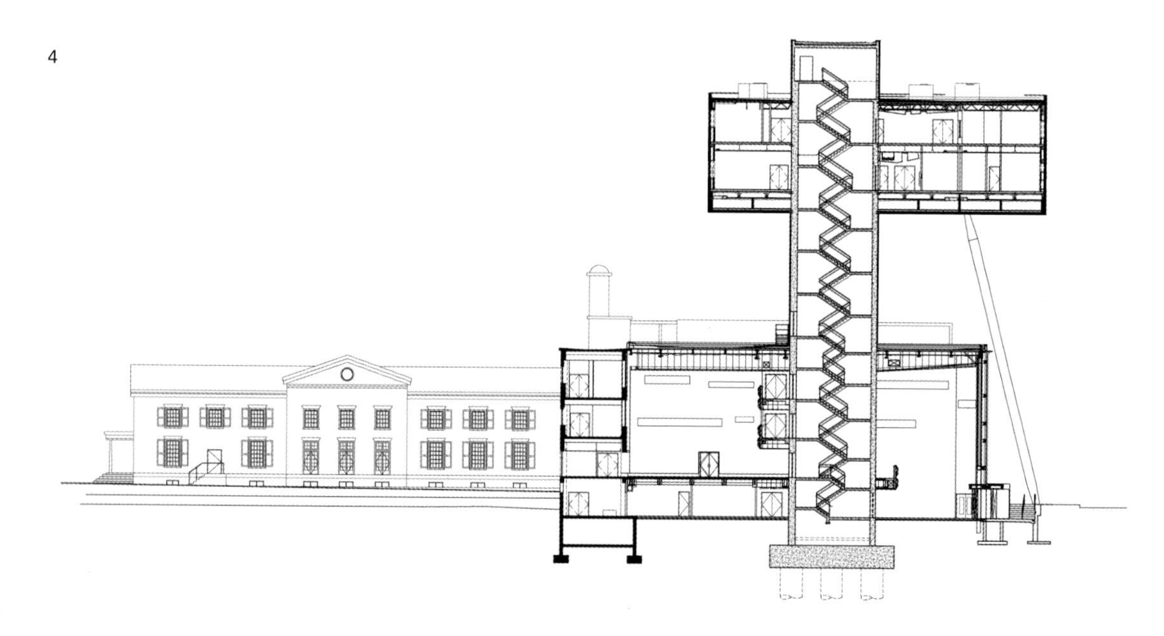

3
Long section, showing how the concrete core provides a vertical link and lateral stability to the elevated 'tabletop' extension

4
Cross-section, showing the extent of the cantilevered frame

5

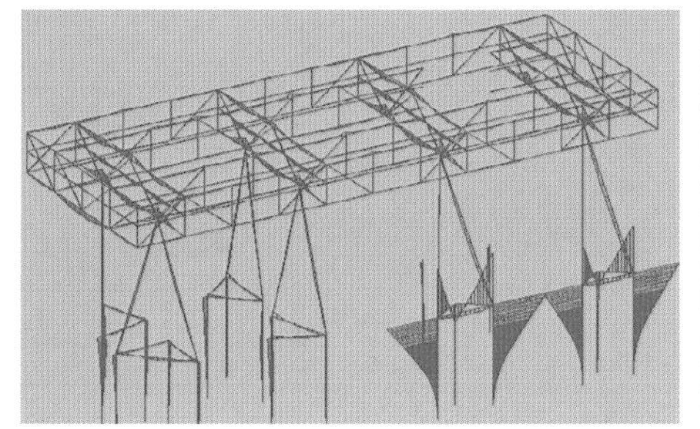

6

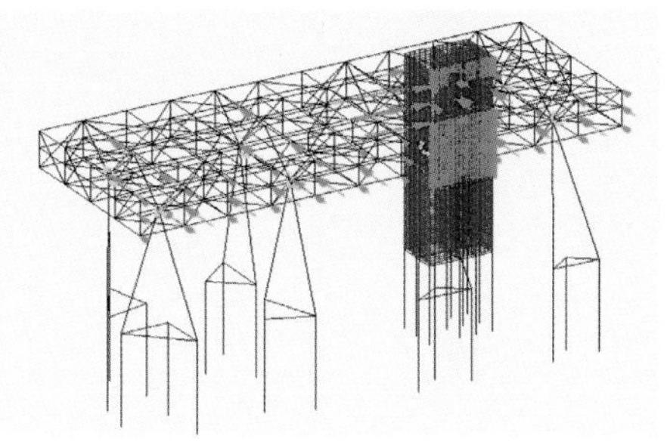

7

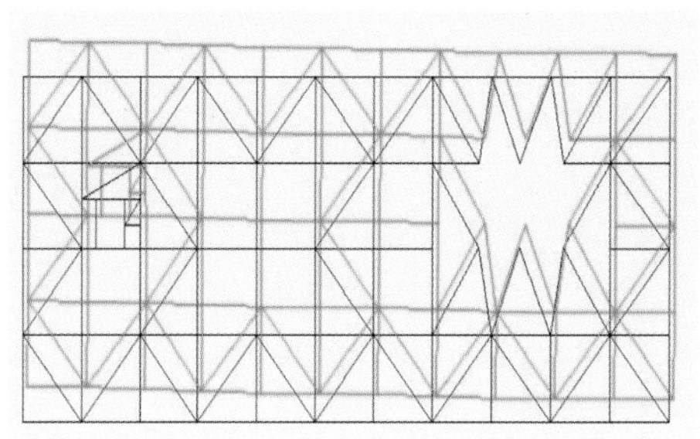

5
Structural diagram, illustrating bending-moment effects and dead (static) load

6
Structural diagram, illustrating wind load in east–west direction

7
Structural diagram, illustrating wind deflection of steel structure at level five

8

8
Steel-frame construction built around the concrete lift/ stair core, with 8 out of the 12 final columns in place

9
Placement of the final two pairs of leg supports. Note the blue-painted steel-leg armatures used to hold the legs in the correct position during construction

10
Detail of double-leg connections to the underside of the steel 'tabletop' structure, with stiffening plates welded to the web of the universal beam

11
View of 29-metre-long steel legs at the fabrication shop, showing the specially rolled, tapered, welded end sections

12
Details of the steel-leg base connection

9

10

11

12

4.4.2
Atlas Building

Structural description
Reinforced pre-cast concrete exoskeleton with steel box beams

Location
Wageningen, Netherlands

Completion date
2005

Plan dimensions
44m long x 44m wide

Height
26m

Architects
Rafael Viñoly (b. 1944) (Rafael Viñoly Architects) with Van den Oever, Zaaijer & Partners Architecten

Engineer
Pieters Bouwtechniek B.V.

The idea of an exoskeleton, and that the structural function of a building could be purposely made visible, is not a new one; the Atlas Building is an excellent recent example of this genre, which notably includes Piano and Rogers' Centre Pompidou and the more integrated diagrid of Norman Foster's Swiss Re 'Gherkin' building. This new seven-storey office and laboratory for Wageningen University is part of the university's move to a new campus in De Born, north of Wageningen. The outer frame is constructed from large double-diamond pre-cast concrete elements measuring 7.2 metres long and 3.6 metres high, with the reinforced concrete elements measuring 400 millimetres wide and tapering to 380 millimetres at the front edge. Cast into the centre of each of the pre-cast components is a steel plate, which picks up one of the specially fabricated steel box beams, and spans across to an internal column to the edge of the atrium. The connections with steel beams and the pre-cast concrete exoskeleton are carefully controlled with slotted-hole and pin connections, ensuring that only vertical load (and no lateral differential movement) is transferred to the frame; two internal concrete cores are designed to resist lateral loading.

The plan of the building is that of a 'square doughnut', and the structural arrangement is such that there are no columns in the open floor space. The pre-cast concrete units are fixed together using a simple keyed joint, and are held in place with steel dowels and chemical fixant. At each floor level, 50-millimetre-diameter steel tension rods are cast into the pre-cast units. The former resist any problematic shear loads caused by thermal expansion of individual units.

The high-quality pre-cast finish of the double-diamond external framework components was achieved with reusable steel formwork and self-compacting concrete. Titanium dioxide, an ingredient more commonly utilized in house paint and toothpaste, was used as an admixture to help whiten the concrete and inhibit mould growth. Recent trials in the Dutch city of Hengelo have also seen titanium dioxide being experimentally used as photocatalytic coating on concrete, which in sunlight will metabolize harmful nitrogen oxides contained in vehicle exhausts into more benign nitrates. The building façades are virtually identical except for cutaway sections at ground level on two sides for access doors and the main entrance, which is a two-storey hexagonal void punched through the latticework of the north façade. A 90-metre-long steel entrance bridge leads you into the building.

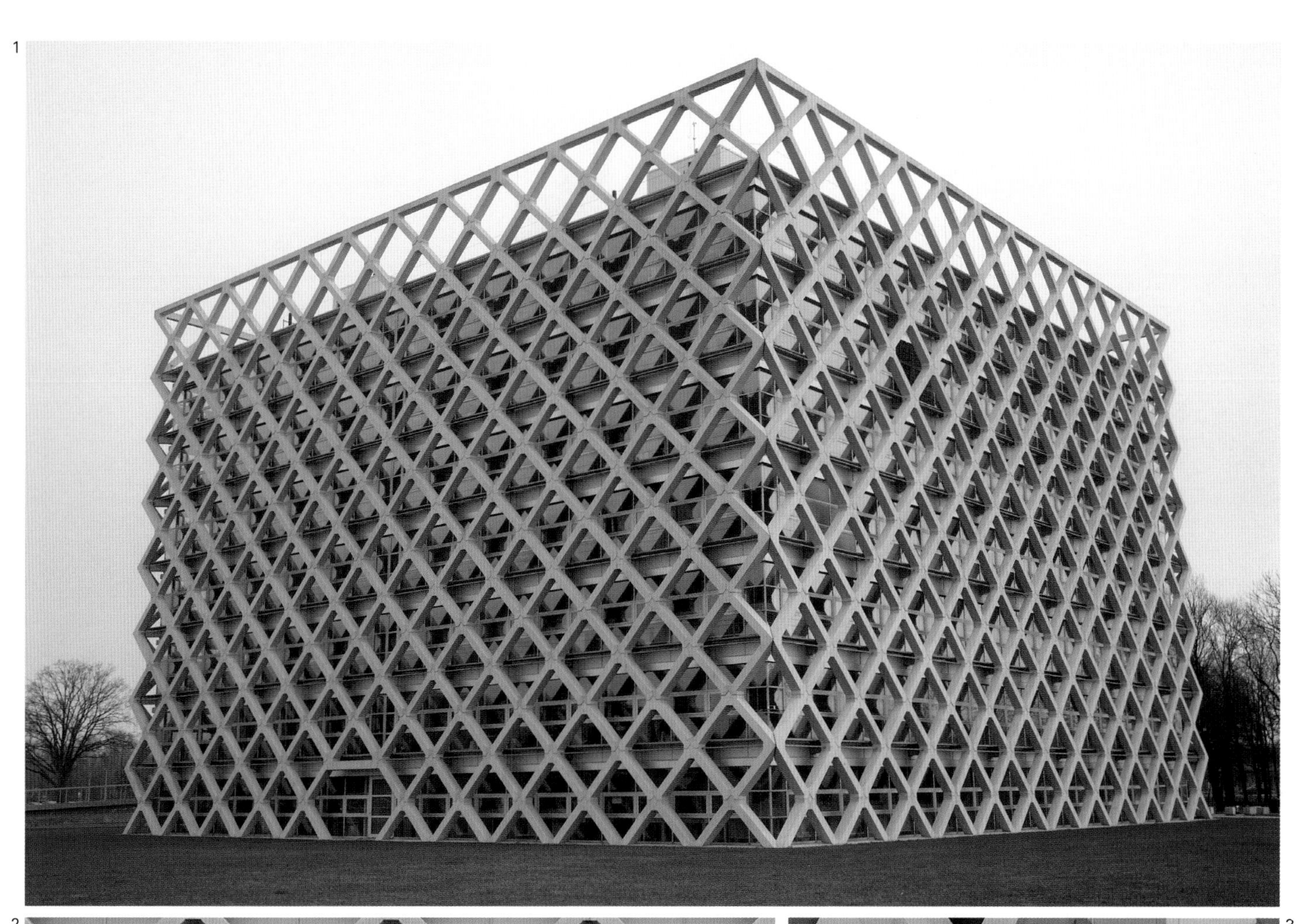

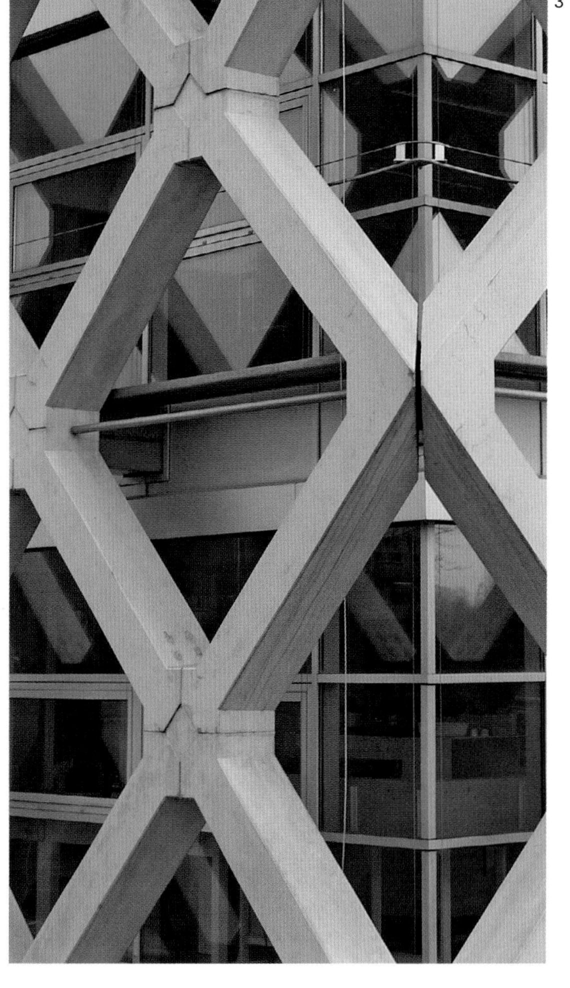

1
South-facing façade of the Atlas Building, with concrete exoskeleton wrapped around and supporting the building

2
Façade detail

3
Corner detail, showing the steel 'internal beam' connection and the horizontal steel tension rods

4.4.3
'Het Gebouw' (The Building)

Structural description
Double (balanced) cantilevered steel tube

Location
Leidsche Rijn, Utrecht, Netherlands

Plan dimensions
27.3m long x 3.9m wide (each block)

Architects
Stanley Brouwn (b. 1935) and Bertus Mulder (b. 1929)

Engineers
Pieters Bouwtechniek B.V.

Completion date
2006

Height
7.8m

This temporary exhibition space is a collaboration between the Dutch conceptual artist Stanley Brouwn and architect Bertus Mulder, known for his restoration of the Rietveld-Schröder House and recent reconstruction of the Rietveld Pavilion at the Kröller-Müller Museum. 'Het Gebouw' (The Building) plays a neat structural game with a square-section prismatic slab perched at a 90-degree rotation atop its close relation, creating a balanced cantilever 11.7 metres long at its greatest extent. Stanley Brouwn is one of Holland's most celebrated artists, and is best known for his conceptual artworks in relation to walking and feet. In a notable series of works from 1960 to 1964, entitled *this way brouwn*, the artist stopped passers-by and asked them to draw directions from a to b. In 1960, Brouwn also documented all the shoe shops in Amsterdam and began to make a series of measured walks. Interestingly, he measured these walks in the Stanley Brouwn Foot (SB foot), which was based on the length of his own foot. One SB foot measures 260 millimetres, and the design of the Het Gebouw pavilion is based upon this module. The length of each block is 27.3 metres (105 SB feet) and the cross-section of each block measures 3.9 x 3.9 metres (15 x 15 SB feet). The building is subdivided into a 5 SB-foot grid, which is clearly visible and helps to organize the building components. The structural challenge was to create a rigid upper level with two identical cantilevers of 11.7 metres (45 SB feet). The structure is fabricated from small hot-rolled steel sections, with bracing provided by diagonal steel rods. Where the two blocks meet, the steel sections are considerably enlarged and moment connections provided with stiffened corner plates. The structure is built using primarily bolted connections and was originally designed to be demounted and relocated.

Het Gebouw sits on the edge of Leidsche Rijn, the site of a new residential development for 80,000 people west of Utrecht, and the pavilion is adjacent to a large geodesic dome constructed from cardboard tubes and designed by Japanese architect Shigeru Ban. 'Het Gebouw' and Ban's 'Paper Dome' were both built as cultural buildings, with Het Gebouw hosting regular art exhibitions and the Paper Dome operating as a community theatre. These projects, commissioned by Bureau Beyond, were to act as magnets and a focus for future developments – once again extending the built-environment frontier, and reminding us why the Netherlands is one of Europe's most densely populated countries. Het Gebouw was originally commissioned for five years, but the building's success as what architect Bertus Mulder describes as 'An autonomous work of art', and increasingly as a local landmark, has persuaded the municipal authorities to retain the structure. However, owing to major construction work in the vicinity the local ground level is being raised by 1.1 metres and as a consequence Het Gebouw will also be raised; Bertus Mulder explained that the building will not have to be disassembled, but can be lifted as a single entity and refixed to a modified and elevated foundation.

Two sections in each block of the building are glazed both sides with an entrance door centrally located in one of the glazed panels, all coordinated with Brouwn's dimensional system. The gallery curator explained that during a recent exhibition-opening party, a large crowd of children and parents had caused noticeable movement in the cantilevered ends: a not unpleasant but slightly unnerving experience. In actuality, Mulder explained, 200 people in one end of this small building would still not be cause for (structural) concern, but it is difficult to see how they would all fit in. As well as an enigmatic work of art and architecture, Het Gebouw is an excellent structural model that illustrates the performative possibilities of simple materials cleverly arranged.

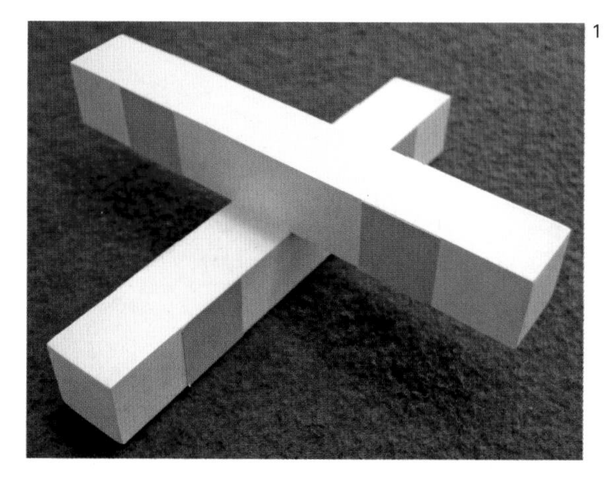

1
Artist Stanley Brouwn's original model

2

3

4

5

2, 3
Het Gebouw and the delicate structural balancing act

4
Beneath one of the gravity-defying cantilevers

5
The steel skeletal framework. Note the diagonal tensile-steel rods in the walls of the upper structure and the heavier, steel SHS elements used in the central (connecting) section

4.4.4
Hemeroscopium House

Structural description
Helical cantilever

Location
Las Rozas, Madrid, Spain

Completion date
2008

Floor area
400m^2

Height
9m

Architects
Antón García-Abril (b. 1969), Elena Pérez, Débora Mesa, Marina Otero, Ricardo Sanz and Jorge Consuegra, Ensamble Studio

Technical architect
Javier Cuesta

Contractor
Materia Inorgánica

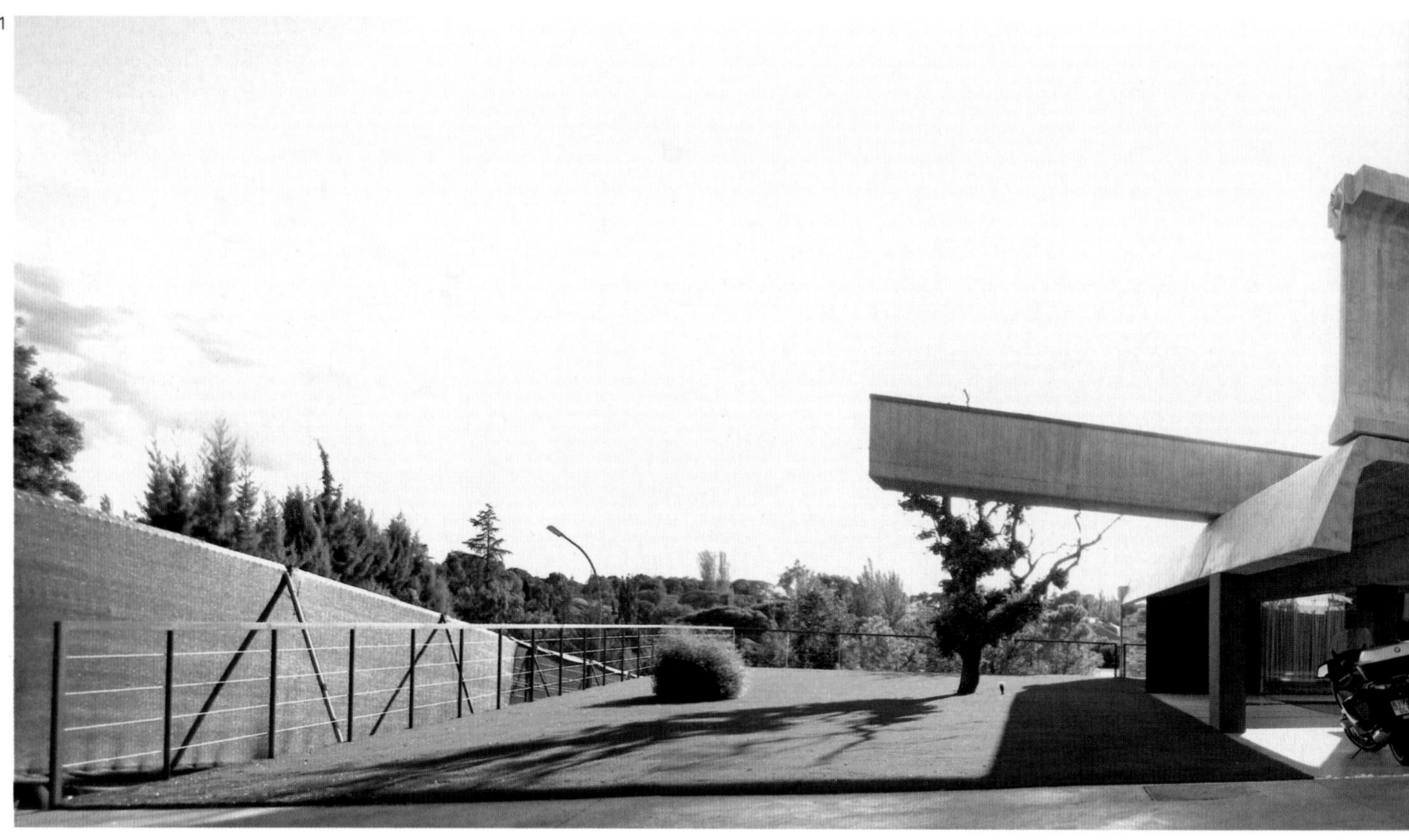

1
Eastern elevation of the Hemeroscopium House, showing granite counterweight

As a structural diagram, a construction sequence and as a set of constructional elements, the Hemeroscopium House is an exceedingly elegant pedagogic tool. Components include a warren truss, a vierendeel truss and three forms and sizes of prefabricated reinforced-concrete beams. This project for a private residence northwest of Madrid also employs a 20-tonne rough-hewn granite boulder as a kind of anchor and counterweight balanced atop the structure, without which we are assured there would be no structure at all. In a recent lecture in London, architect Antón García-Abril explained that it is the 'gravitational traces that make the space'.[1] The complex engineering and design for the project took one year, but the structural frames took a mere seven days to assemble. The structure for the house consists of seven key elements, which are stacked up using a helicoidal arrangement. The first element is the stable and heaviest 'mother beam', which is 22 metres long and 2.65 metres high and weighs a not inconsiderable 59 tonnes. This concrete I-beam is prefabricated off-site and uses specially designed pre-tensioned steel reinforcement to achieve its desired strength. The second element is an inverted U-shaped beam of 22 metres, which picks up another massive concrete I-beam at its cantilevered end and also, at its midpoint, a U-shaped concrete beam that has reinforced glazed ends, is filled with water and acts as an

elevated, linear swimming pool. That this 21-metre-long pool contains 25 tonnes of water only seeks to reinforce the complex network of structural interdependencies. As the low-slung structural helicoids rise, a transparency of beam elements is introduced using steel to create the fifth and sixth spanning elements – a steel vierendeel and warren truss, respectively – and the seventh and final beam is another concrete I-beam, upon the end of which sits the granite counterweight drilled through and bolted to the beam. This counterweight allows the last beam, which is balanced atop the water-filled beam, to cantilever at its other end and support the steel warren truss. These complex structural relationships between the elements and the structure in its entirety are, however, explicitly illustrated. There has been no attempt to hide or obfuscate structural actions, with this single house able to tell a number of stories, both structural and spatial.

1 García-Abril, A., 'Stones and Beams', lecture given at the Architectural Association School of Architecture, 2 March 2011

2

3

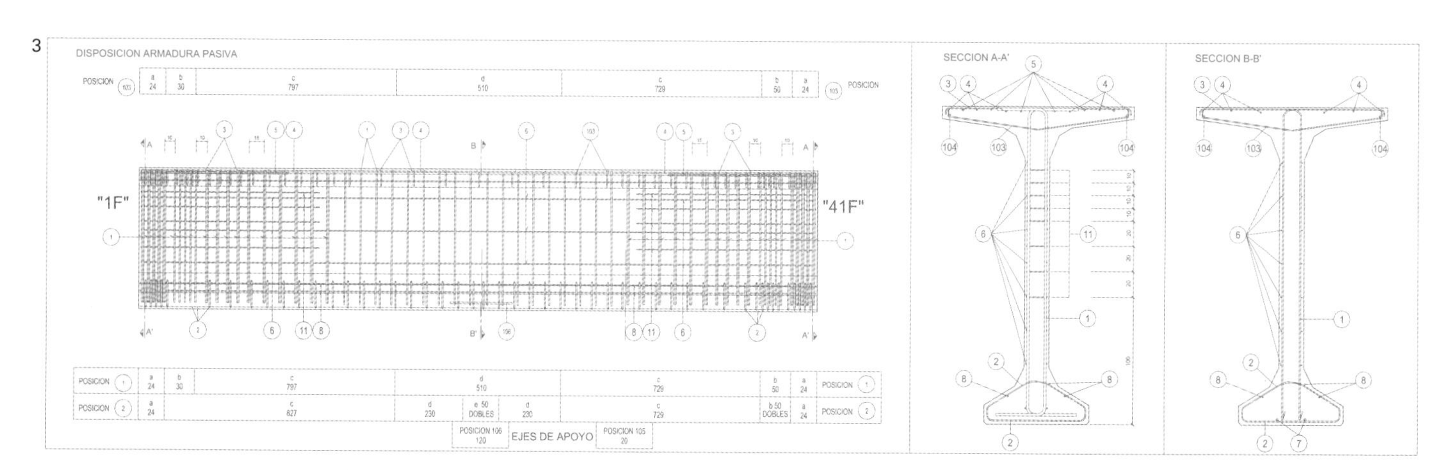
DISPOSICION ARMADURA PASIVA
"1F"
"41F"
EJES DE APOYO
SECCION A-A'
SECCION B-B'

2
South elevation, with cantilevered linear pool

3
Fabrication drawing of concrete beam no.3, showing the distribution of steel reinforcement

4
Axonometric drawing showing structural logic and sequential assembly

4

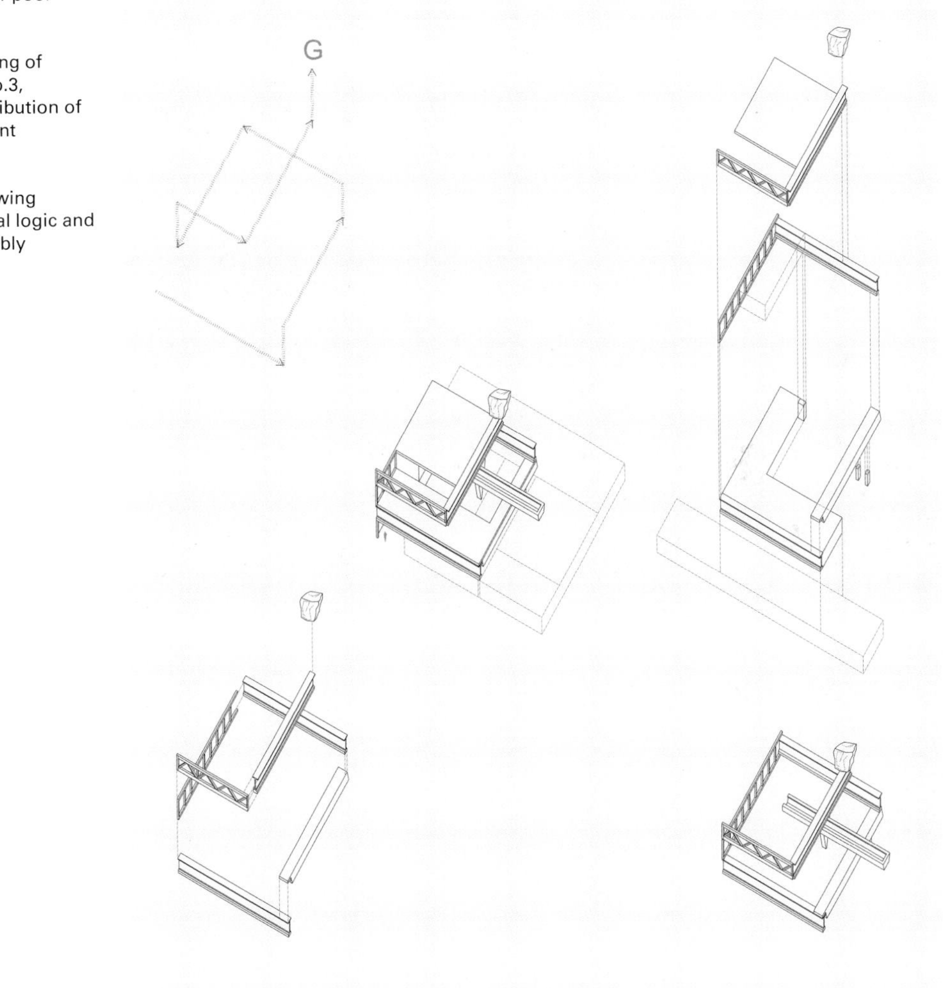

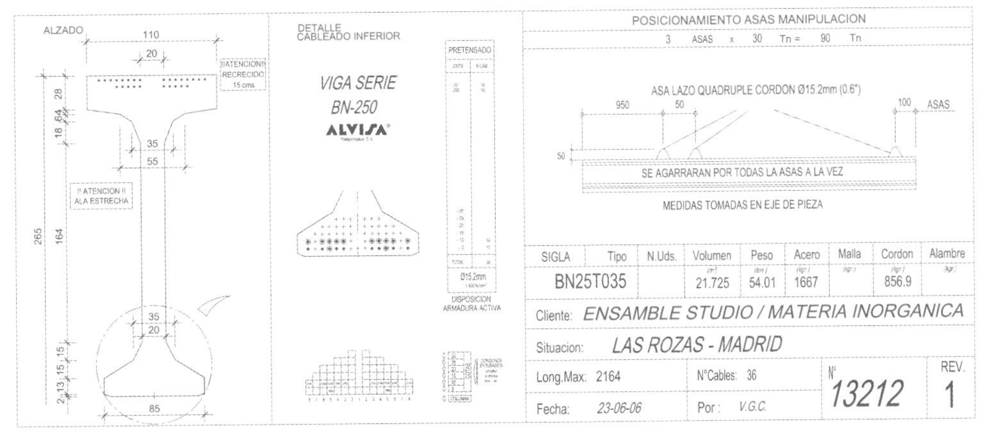

4.4.5
Kanagawa Institute of Technology (KAIT) Workshop/Table

Structural description
Post-tensioned, structurally optimized steel frame

Location
Kanagawa, Japan

Completion date
2008

Floor area
1,989m²

Height
5.052m

Architect
Junya Ishigami (b. 1974) (Junya Ishigami + Associates)

Engineers
Konishi Structural Engineers

Contractor
Kajima Corporation

Software
Tomonaga Tokuyama

Though not immediately obvious to the eye, there are two different types of columns in the structure, the verticals (those bearing vertical forces), and the horizontals (those bearing (or resisting) horizontal forces). I wanted to make the columns as slender as possible, and assigning the forces was more effective than trying to make every column bear both. I didn't want just any sort of slender columns.[1]
Junya Ishigami

Junya Ishigami is a young architect who does not seem to be afraid of producing wonderful pieces of architecture and design while simultaneously employing the creative potential of engineering properties and dynamics. If Ishigami's Table project of 2005 is both engineering set piece and conjuring act – which, based on our engineering preconceptions, appears to defy gravity – then his workshop for the Kanagawa Institute of Technology (KAIT) is a complete work of architecture and engineering.

Commissioned as part of the university's redevelopment of its campus, this 2,000-square-metre single-storey structure was designed as an open-access studio facility, available for students to undertake project work in a range of different media. The architect conceived of the building as a stroll through the woods that cleverly delineates the one-room environment into ambiguous domains through column densities and disposition 'in a way that gives no hint that any rules or plan for their placement exist'.[2] The plan of the building is a slightly skewed square; a single roof plane is held aloft by 305 columns, each unique in its cross-section sizing and orientation. The building is glazed on all sides, and the deep plan lit throughout with strips of rooflights.

Although this project is visually stunning, in that a lightness and transparency is maintained with numerous yet startlingly slender columns, this is only half the story. The engineering of this structure is an extremely precise and yet unorthodox mix of parametrically defined precision and a radically innovative hierarchy of structural 'action designation'. Ishigami creates two 'classes' of column, ostensibly for two different functions: one set of columns to carry the load of the I-beam roof grid and one set to resist lateral movement, which the architect calls the vertical and horizontal columns. Where Ishigami has been extremely skilful is in creating a lightweight forest of columns where the specific structural function of any given one cannot be identified. The supports are all differently sized and vary from column no. 240 (a compact 80 x 56-millimetre solid steel section) to column no. 277 (160 x 16-millimetre steel flat), with all of them specifically orientated at angles down to a tolerance of one decimal place. The construction of KAIT was critical in maintaining Ishigami's aim of creating an even treatment of all column connections.

The columns are erected using two different processes according to their type. For the verticals, the bases are joined to independent foundations, with steel I-beams placed across the top ends. Pin joints are used to attach the verticals to the beams. The detail of these pins is ultimately concealed in order to match the welded detail of the horizontals' top ends. In order to keep the horizontals (the lateral-resisting columns) slender and prevent their own weight acting on them as a vertical force, they were suspended from the roof beams. After the verticals were joined to the beams, the horizontals were inserted with a crane from above the beams and fixed. The horizontals are not intended to bear snow loads and other vertical forces, so the initial design was to keep their connection with the floor vertically loose to avoid potential buckling from snow load. The problem with making loose holes and inserting the horizontals into them was that you would make visible details that didn't match the corresponding details of the verticals at the floor–column connection. This outcome was unacceptable to Ishigami, so he used another approach: before fixing the horizontals to the beams the roof was pre-emptively loaded with weights equal to the snow load, and then the columns were fixed. When the temporary loading is removed, the horizontals (lateral-load columns) are put into tension, thus preventing bending if the structure is snow loaded. The process maintains the required ambiguity of structural function that Ishigami required while creating a new structural type – or at least a new structural approach, achieved through very detailed analysis using software developed by his firm.

Junya Ishigami belongs to a long line of designers for whom the structural strategy, logic and material use of any given design are co-dependent with the programmatic ambitions of a design project. His Table project, which has been exhibited in Basel, London, Tokyo and Venice, is worth mentioning in relation to the general themes of this book for its structural audacity and creative 'reverse engineering'. A table surface (9.5 metres long x 2.6 metres wide x 1.1 metres high) of 3-millimetre-thick steel is held by four legs, one located at each corner. That the table can support its own material weight over this span is seemingly impossible; that it can support everyday objects such as fruit bowls and vases seems illusory at the very least. What Ishigami has done has pre-rolled the top of the table, like the spring of a clockwork mechanism, and the tabletop is only bought level when unrolled and carefully loaded with precisely placed and weighted objects. The table is so delicately balanced and structurally optimized as to slowly ripple to the touch. As is shown by the KAIT project, Junya Ishigami's innovations are both structurally inventive and polemically rich and provide clues to hitherto unimagined design solutions for a new generation of architects and engineers.

1, 2 Ishigami, J., Project information provided by the office of Junya Ishigami & Associates, 2011

4.4.5
Kanagawa Institute of Technology (KAIT) Workshop/Table

1

3

4

1
Plan drawing, showing the unusual layout of the 305 columns, which are indicated by dots

2
Diagram showing the roof-beam structure and the two designated 'classes' of column

3
KAIT under construction, showing the separate column-cluster foundations

4
View of roof structure during construction, with red prime-painted steel sections to left of picture temporarily in place to replicate snow loading

5
Section drawing through the edge of the KAIT building

2

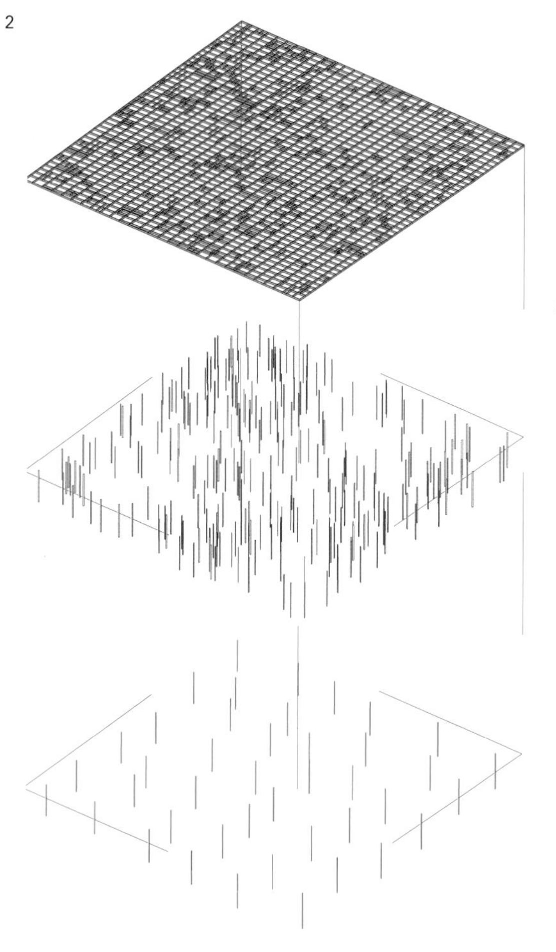

5

6
Exterior view of completed building

7
Interior view of finished project before occupation

8
Architect's drawing of the Table project, with locations of table objects and their weight

9
Finite Element Analysis (FEA) diagram of the tabletop

10
Elevation drawing of the Table in its 'deployed' and 'un-deployed' (rolled-up) state

11
Factory photograph showing the steel tabletop being rolled (prestressed)

12
The 'gravity-defying' finished, fully laden Table

6

7

8

9

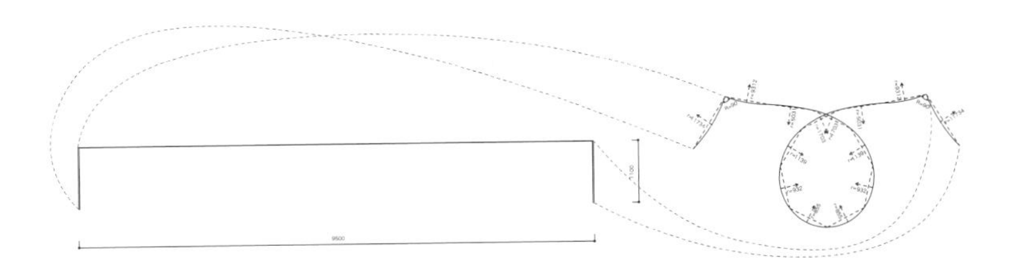

10

11

12

4.4.6
Meads Reach Footbridge

Structural description
Portal-frame profile with a stainless-steel stressed skin

Location
Bristol, England

Completion date
2008

Length
55m

Architect
Niall McLaughlin (b. 1962)

Engineer
Timothy Lucas (Price & Myers)

Artist
Martin Richman

The art of structure is how and where to put the holes.[1]
Robert Le Ricolais

The brief from the client for an 'invisible' walkway for pedestrians and cyclists over Bristol's floating harbour was developed by architect Niall McLaughlin, engineers Price & Myers (Geometrics Group) and the light artist Martin Richman. The ambition of 'invisibility' led the design team to look at a perforated surface that would not have to be lit at night but could be a source of illumination itself, emitting light through a distribution of holes.

The structural form of the bridge is that of a four-legged portal frame with flexible, pinned base connections at each end. The span is achieved by using the torsion-box principle of a plane wing, creating a stressed-skin structure made entirely from grade 2205 stainless steel. The bridge is formed from a series of perforated stainless-steel ribs, connected to a thin-plate perforated stainless-steel spine element; the ribs are also connected by intermediate longitudinal sheet steel struts and internal cross-bracing elements inside the deck. This relatively lightweight framework is then wrapped in 6-millimetre stainless-steel perforated sheets, which are welded to the subframe assembly (this is a fully welded structure). The depth of the balustrades is effectively forming the bridge's spanning capacity, with the underside of the structure providing lateral stiffness. The 'walkable' deck of the bridge is the only element that is not welded, and it is formed from a series of removable textured and perforated stainless-steel panels. These panels allow access to the lighting battens fixed inside the bridge. The bottom edge profile of the bridge is formed from a solid stainless-steel rod, which helps to resist tensile forces.

The perforations that cover the bridge are interesting in a number of ways; primarily employed so as to allow the bridge to luminesce in darkness, putting holes in a bridge is also structurally intriguing. Although there is a risk that you structurally weaken the bridge, you are also removing material and thus lightening the static 'dead' load, which is structurally beneficial. The size of the perforations varies from a diameter of 10 millimetres to a maximum of 40 millimetres. The holes are positioned at regular centres, with their diameter locally determined from a Finite Element Analysis (FEA) of a stressed-skin unpunctured model. The engineers managed to link their structural data map to a spreadsheet, which produced a series of numerical maps with varying perforation diameters detailed. This

1
Elevation of 'portal' bridge. The portal design creates rigid connections at the haunches of the bridge, while the pinned base connections allow for the thermal expansion and live loading of the structure

2
Exploded view, showing construction elements

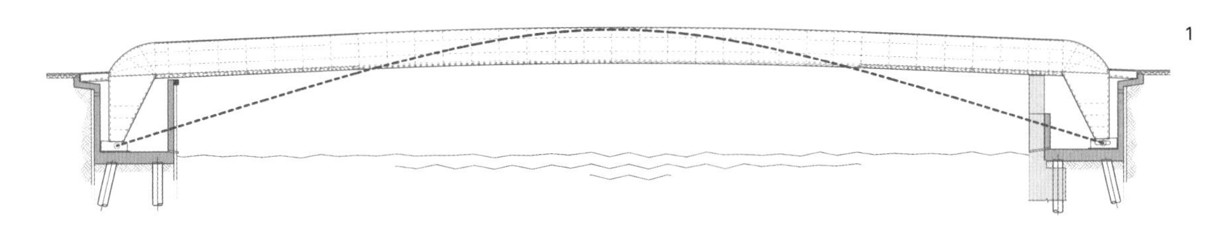

1

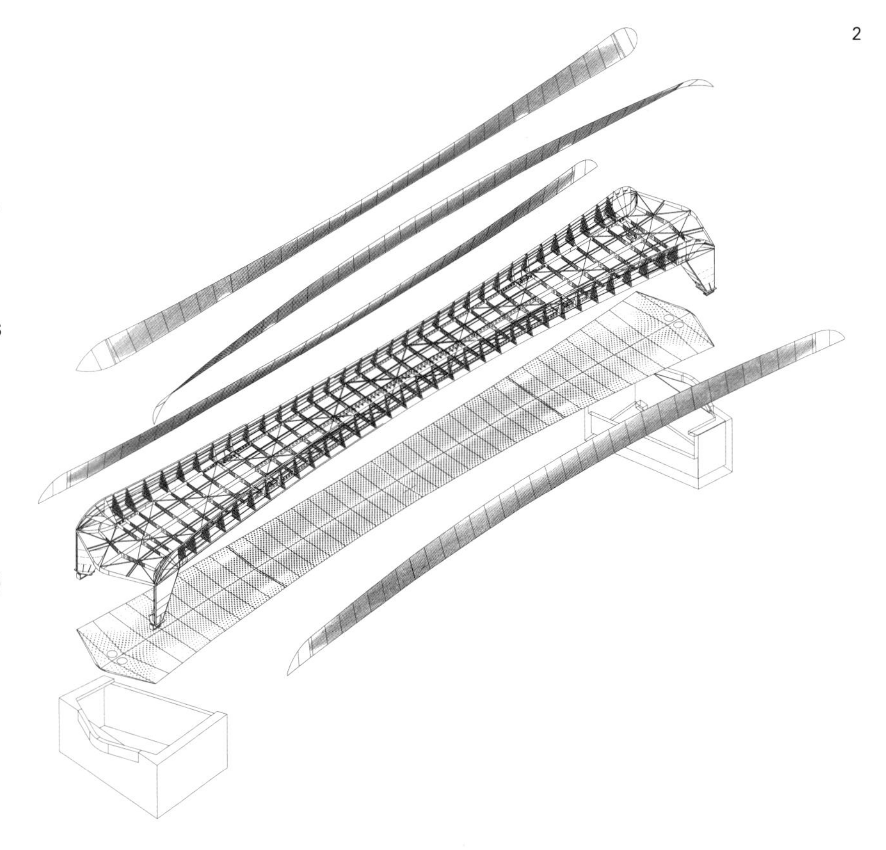

2

information could be sent direct to the CNC plasma cutters that were cutting the steel sheets for the bridge. In areas of high stress distribution, such as the haunches of the 'portal' bridge legs, the holes decrease in size, and sometimes there are no holes at all. Niall McLaughlin has said, 'the pattern of holes becomes a stress map of the work the bridge has to do to cross the river'.[2] In all, there are 55,000 perforations. The bridge was pre-assembled in sections, which were welded together on a vacant plot adjacent to the final location, and the 75-tonne bridge was lifted whole by a mobile crane into its final position.

The bridge links the harbour to the city centre, and has received awards from both the Royal Institute of British Architects (RIBA) and the Institution of Structural Engineers.

1 Quoted in Sandaker, B. N., *On Span and Space: Exploring Structures in Architecture*, Oxford: Routledge, 2008, p. 71
2 Spring, M., *Meads Reach footbridge, Bristol*, PropertyWeek.com, 23 July 2010

3

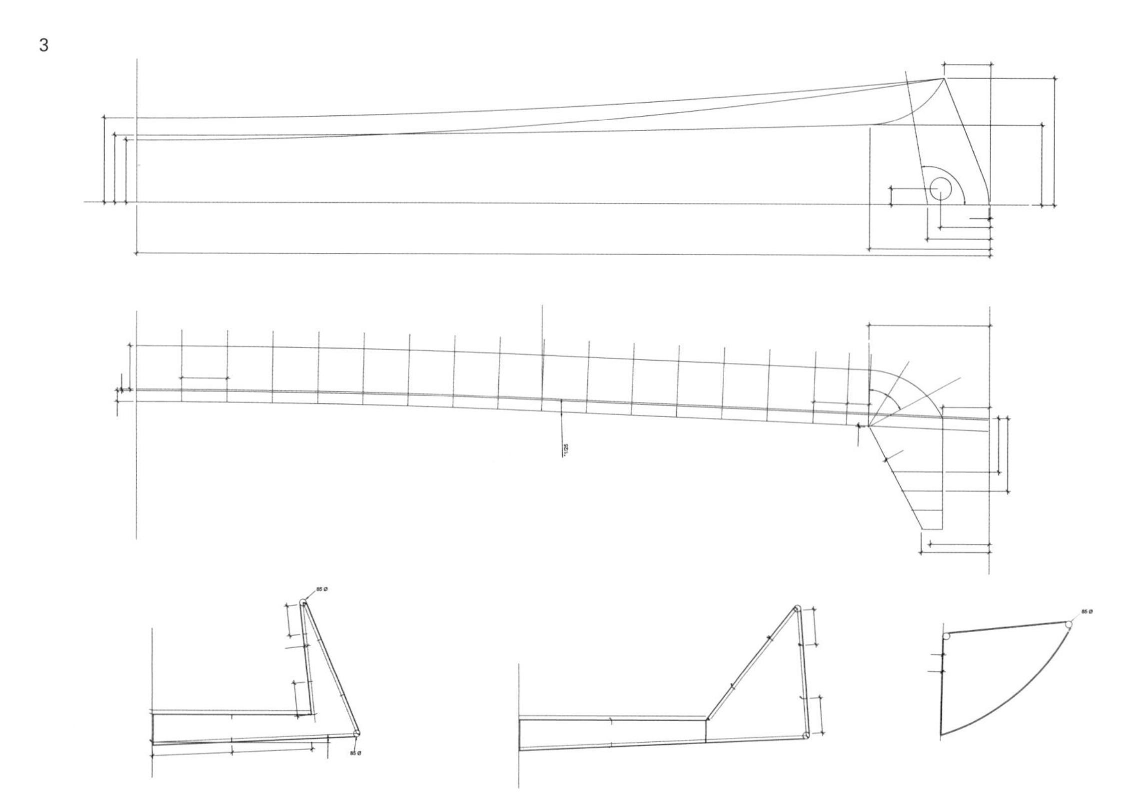

3
Detail drawings: plan, elevation and rib details

4
Visualization of stress distribution through Finite Element Analysis (FEA)

5
A sample of the spreadsheet used to generate the machine code for automated CNC laser cutting of the perforations

6
Developable surfaces: the geometrically complex surfaces of the structure were carefully modelled to allow them to be developed from flat sheets, for ease of fabrication

7
3D model of stainless-steel component with variably sized hole cut-outs

8
3D model

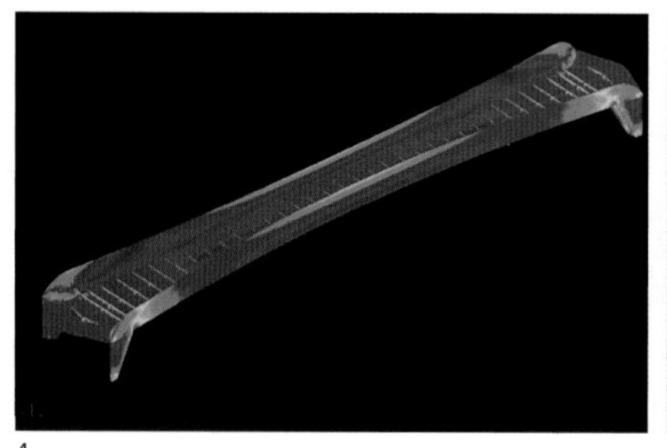

4

5

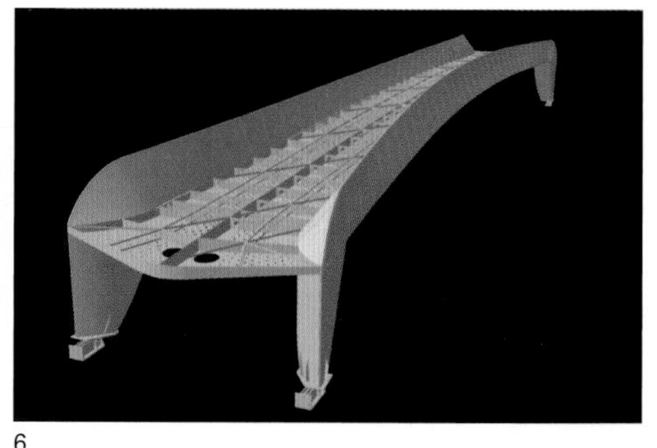

6

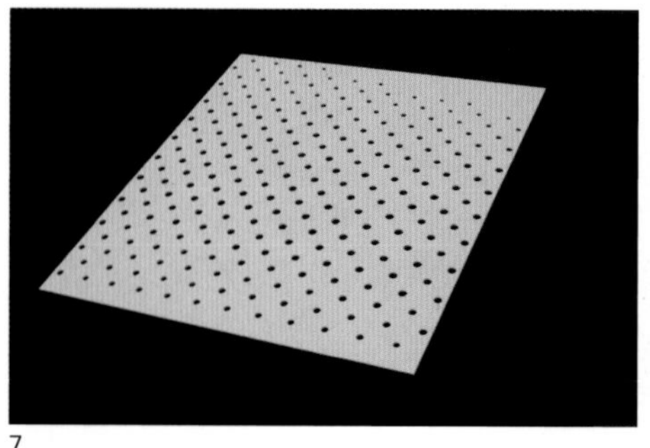

7

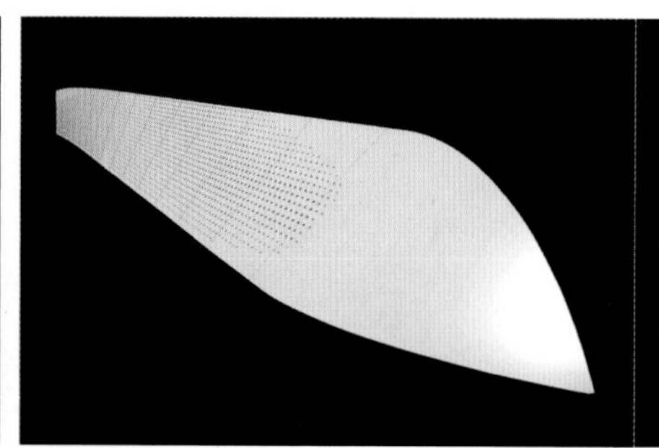

8

9
Fabrication of bridge in stainless steel, showing ribs and spine elements

10
Lifting the bridge (whole) into position

11
The Meads Reach Footbridge is illuminated, so that the inner ribbed structure is revealed at night

12, 13
Detail of finished bridge

4.4.7 Pompidou-Metz

Structural description
Timber gridshell roof structure

Location
Metz, Lorraine, France

Completion date
2010

Plan dimensions
Hexagonal roof 90m wide – 8,000m²

Floor area
10,700m²

Architects
Shigeru Ban (b. 1957), Jean de Gastines, Philip Gumuchdjian

Engineers
Terrell Group

Contractor
Demathieu & Bard

Production software
Design to Production

Specialist timber fabricator
Holzbau Amann

Fabric membrane
Taiyo Europe

I bought this hat 10 years ago in Paris, but it's the same you see everywhere in Asia, usually worn by field workers. It has a bamboo structure, a layer of insulation, and oil paper as waterproofing. The building has the same fundamental elements, including the hexagonal weave pattern.[1]
Shigeru Ban

Pompidou-Metz, a new outpost of the eponymous Paris-based parent institution, is an exhibition space for visual art with a restaurant, shop and auditorium. The three main gallery spaces are 80-metre-long rectangular tubes stacked on top of each other with picture windows at each end. A 77-metre-high concrete-and-steel tower connects the gallery spaces, and the entire structure is wrapped in a fabric-clad hexagonal timber-gridshell structure.

The roof of the new Pompidou-Metz was inspired in part by the woven canework of a Chinese hat that architect Shigeru Ban found in a Paris market. The roof, hexagonal in plan, is a giant, triaxial, woven, double-layered timber gridshell with a three-way parallel grid of 2.9-metre modules. The structure consists of 650 tonnes of glue-laminated timber elements, prefabricated in a German factory and assembled on site. The majority of these elements are glulam planks 440 millimetres wide, 140 millimetres deep and approximately 15 metres in length. The planks are overlaid in three directions and then a second layer of planks is added with timber blocks between, increasing the depth and thus the structural performance of the assemblage. A tubular concrete-and-steel tower, which contains the vertical circulation and access to the elevated gallery elements, supports the prow of the timber roof 'hat' with a tubular steel ring. Similar rings are also used to form four openings in the roof for the protruding galleries. Interestingly, assembly of the timber gridshell was started from its highest point, and by using scaffold support towers the timber framework was built outwards from this central tower to the edge beams. The edge beams themselves are also glulam timber, but with a considerably deeper section than that of the roof; they work as simple two-dimensional arch structures, minimizing the number of edge supports to six: one for each apex of the hexagon. The edge supports are formed by pulling the gridshell down through the horizontal plane to form six three-dimensional latticework columns, set back from the edge of the structure. This complex timber gridshell spans up to 40 metres.

Although hexagonal in plan this is not a symmetrical surface; the geometry of this timber grid is pulled up and down through the horizontal plane, utilizing both synclastic and anticlastic curvature to provide stiffness. The tighter radii of the lattice columns provide excellent structural stiffness and

1

1
View of a virtually complete Pompidou-Metz at night, with the timber gridshell clearly visible through the PTFE fabric skin

resistance to wind loads. The structure underwent rigorous wind-tunnel testing at Nantes' CSTB (Centre Scientifique et Technique du Bâtiment), as well as testing for snow loadings and subsequent internal climatic effects.

The original structural design of the project was undertaken by Cecil Balmond's specialist engineering studio, Advanced Geometry Unit, at Arup. This early design differed from conventional gridshells in that it employed the use of reciprocal beams fabricated from steel and timber, specifically designed to simplify the connections by co-joining the woven node points in a structural 'sandwich' or lamination. The final realization of the project used the more conventional gridshell system of a double-layered three-way woven timber grid, comprising six layers of glulam timber planks at 400-millimetre offsets with steel bolts connecting the node points. The use of glulam timber, however, made it possible to preform the planks with a specific radius: each was fabricated with a single custom curve along its length, and was then Computer Numerical Control (CNC) milled to introduce a secondary twist (or curvature). It would have been possible to laminate the timber planks in two directions, but for fabrication purposes it was decided to create oversized, single, curved elements and machine the additional curvature. The laminated timber elements are connected end to end using steel plates spliced into the head of each plank and then bolted.

The timber structure is covered with a waterproof membrane made from fibreglass and Teflon (PTFE or polytetrafluoroethylene). The PTFE is cut from flat sheet and assembled into panels using pattern-cutting software to precisely mimic the timber form. The membrane is then connected back to the structure using T-section steel elements. The fabric is held 300 millimetres away from the timber structure, to allow for a smooth airflow and prevent condensation.

1 Lang Ho, C., 'Interview: Shigeru Ban' in *Modern Painters*, 28 May 2010, p. 22

4.4.7
Pompidou-Metz

2
Diagram showing the double curvature of the timber 'planks'. The first curvature is created by glue-laminating single curved elements along the length of the member, with additional curvature (or twist) introduced by machining the timber element across the short section

3
Double-curved glulam timber planks being prepared at the factory in Germany. Each plank is approximately 15 metres long

4
Detail of special turnbuckle tool, created to winch the planks together on site

5
Computer model of the timber latticework, with all elements to scale

6
Construction picture, with laminated-timber perimeter edge beams clearly visible and the tubular-steel framings fixed around the protruding galleries

2

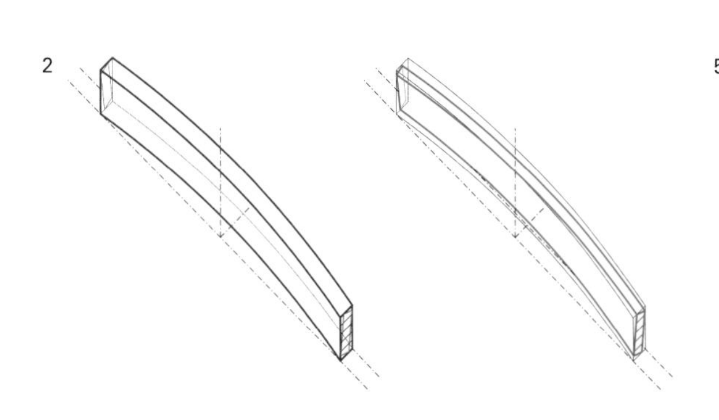

5

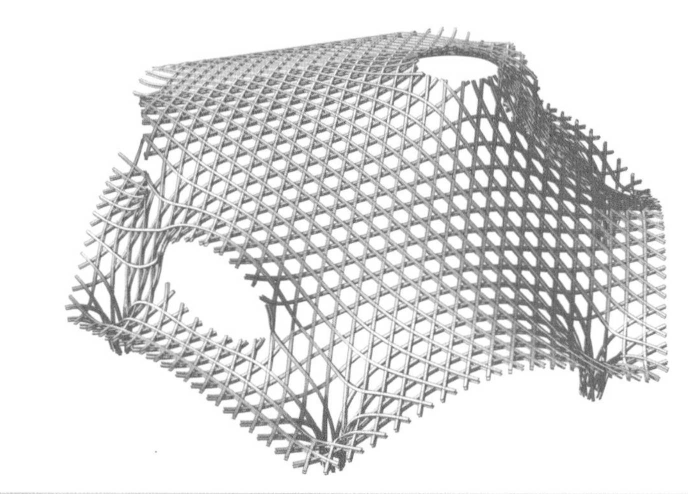

3

4

6

7
Detail of computer model, showing the tight geometry of the lattice leg elements

8
Detail of a timber lattice leg support, showing the steel ring that holds down the PTFE fabric covering

9
Diagram showing the geometry and assembly of the three-way double-layered timber lattice structure

10
Looking from the top of one of the gallery tubes, we can see the second layer of timber planks being laid over the lattice, with spacing blocks shown

11
Interior photograph showing the intersection of the tubular-steel service tower and the apex of the timber roof structure

7

9

8

10

11

4.4.8
Burj Khalifa

Structural description
Buttressed core tower

Location
Dubai, United Arab Emirates

Completion date
2010

Floor area
280,000m^2

Height
828m

Architect and engineer
William F. Baker (b. 1953), Skidmore, Owings and Merrill (Partner in Charge of Structural and Civil Engineering)

Contractor
Samsung/BeSix/Arabtec

Foundation contractor
NASA Multiplex

While the world's tallest building, and indeed the world's tallest man-made structure, is located in the Middle East, the Burj Khalifa is very much a product of North American engineering, and more specifically the high-rise progenitor of Chicago. The location of the tallest 'skyscraper' has been a constantly changing competition that follows economic migrations and has now found its way to Dubai. Chicago-based Skidmore, Owings and Merrill's role in the evolution of the high-rise is significant, with five out of ten of the world's tallest buildings being the work of SOM. In this context, it is important to make reference to the remarkable contribution that SOM engineer Fazlur Khan made to development of new forms of high-rise structural thinking, such as the 'trussed tube' of the John Hancock Center and the 'bundled tube' of the Sears Tower (now renamed the Willis Tower). The legacy of Khan's quiet but significant innovations still resonates in the construction of tall buildings, where material and structural efficiencies are achieved through new geometric configurations and radical rethinkings of engineering orthodoxy.

At 828 metres high, the Burj Khalifa sets a new building-height record, which for economic reasons alone is unlikely to be surpassed any time soon. This predominantly residential block was conceived with a Y-shaped plan, the utility of which delivers increased surface area (and thus vistas for its residents). More important, however, in what is undoubtedly a major engineering achievement, is the increased structural stability that the tapering Y-shaped form affords, employing what William F. Baker describes as a 'buttressed core' structural system. The core, a hexagonal tube that contains all the vertical access, is buttressed at 120-degree intervals by three tapering accommodation wings. Unusually for a building of this immense height, the external form of the structure is asymmetric, which is not a lightly used design conceit but the result of extensive wind-tunnel testing and numerous Computational Fluid Dynamic (CFD) modellings of the tower, which confirmed that tapering the structure and offsetting stepped changes in building width would prevent the consolidation of organized vortex shedding and substantially reduce the wind forces acting on the building. The effects of the wind are also mitigated by the glazing mullions, or 'fins', which SOM have likened to the dimples on a golf ball, 'to create surface turbulence and reduce the lateral drag forces on the building'.[1]

The building is constructed of reinforced concrete – a feat that would have been unimaginable in 1965, when Fazlur Khan had seemingly pushed the limit for high-rise reinforced-concrete design with the 37-storey Brunswick Building in Chicago. New analysis techniques and the refinement of concrete technology have made the Burj project possible. The technical challenges, however, of pumping concrete to such heights over such long distances and in such extreme heat were considerable (UAE temperatures can

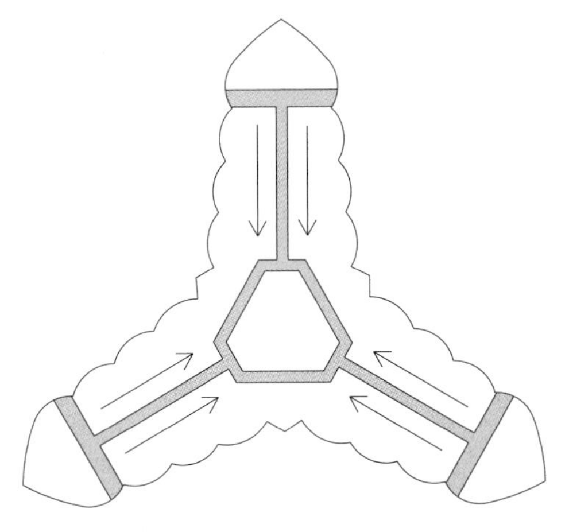

exceed 50˚C). Other technical challenges for a building this large, and with such significant static loads, are the time-dependent changes of concrete shrinkage and creep; over a 30-year period it is predicted that vertical shortening will reduce the overall height of the building by approximately 300 millimetres. This shrinkage and creep also creates changes in the structural performance of reinforced concrete insomuch as it alters the ratio of how much load is taken up by the concrete and how much by the internal steel reinforcing (rebar). It has been estimated that immediately after construction the concrete in the walls and floor at level 135 will support 85 per cent of the load, with the rebar supporting 15 per cent. It is predicted that after 30 years this ratio will have changed to 70:30 per cent, with the rebar taking twice the load it did when the structure was completed.

The building is supported on a solid reinforced-concrete raft, 3.7 metres thick; the material is a C50 Self Compacting Concrete (SCC). This concrete raft is supported by 194 friction piles, each 1.5 metres in diameter and 43 metres long, and each designed for a load capacity of 3,000 tonnes. The groundwater in the site was found to contain high concentrations of chloride and sulphate, which could prove extremely corrosive to the foundations. Several strategies were employed to prevent this potentially harmful corrosion, such as a specially formulated concrete mix, various waterproofing technologies and cathodic protection, which utilizes a titanium mesh beneath the raft along with electricity to repel harmful chemicals.

The construction sequencing of the tower was also vital to the long-term durability of the structure, especially in light of the asymmetrically spiralling layout of the structural setbacks. The superstructure of the tower uses a range of concrete mixes, from C80 to C60 cube strength containing Portland cement and fly ash, and was constructed using a self-climbing (jump form) system, and a mixture of specially designed steel formwork for curved columns and proprietary systems for the concrete decks.

The vast scale of this project is perhaps best illustrated with climate data, which shows that a ground temperature of 46.1˚C is reduced to 38˚C on the 162nd floor at the top of the tower; similarly, there is a 30 per cent reduction in humidity between the top and bottom of the building.

1 Baker, W. F., Mazeika, A., and Pawlikowski, J., 'The Development of Burj Dubai and The New Beijing Poly Plaza' in *Structures Congress 2009: Integrated Design: Everything Matters*, American Society of Civil Engineers, pp. 1–10

1

2

1, 2
The completed Burj Khalifa, currently the world's tallest building: note the asymmetrical setbacks designed to 'confuse' the wind

4

3

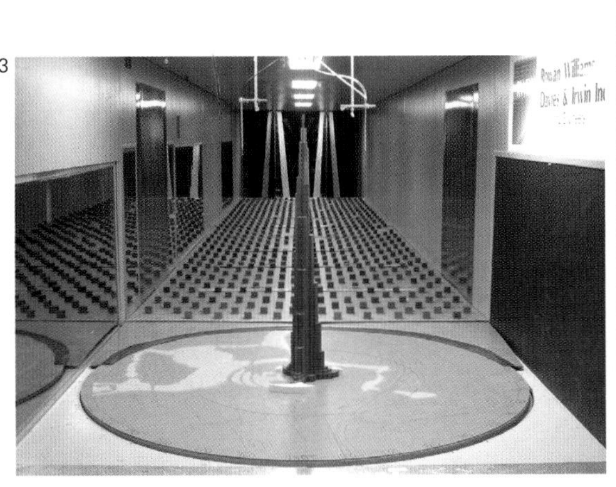

3
Wind-tunnel testing of a 1:500 scale model of the tower. The wind tunnel models contain pressure taps to collect wind data from different areas of the building model

4
Detail of the concrete structure with a four-storey section of cladding in place. The fast-track nature of contemporary construction means that the structural build and finished cladding are programmed simultaneously to allow for internal fit-out

5
The Burj Khalifa under construction with the tripartite plan of the buttressed core visible

5

Further reading and resources

Ackermann, Kurt (et al), *Building for Industry (Industriebau)*, Surrey: Watermark Publications, 1991

Adams, Jonathan, *Columns: Detail in Building*, London: Academy Editions, 1998

Addis, William, *Creativity and Innovation: The Structural Engineer's Contribution to Design*, Oxford: Architectural Press, 2001

Anderson, Stanford (ed.). *Eladio Dieste, Innovation in Structural Art*, New York: Princeton Architectural Press, 2004

Bechthold, Martin, *Innovative Surface Structures Technologies and Applications*, Oxford: Taylor & Francis, 2008

Beukers, Adriaan, *Lightness: the inevitable renaissance of minimum energy structures*, Rotterdam: 010, 1998

Bill, Max, *Robert Maillart: Bridges and Constructions*, London: Pall Mall Press, 1969

Billington, David P., *The Art of Structural Design: A Swiss Legacy*, New Haven: Yale University Press, 2003

Blaser, Werner, *Mies van der Rohe*, London: Thames and Hudson, 1972

Blockley, D., *The New Penguin Dictionary of Civil Engineering*, London: Penguin, 2009

Boaga, Giorgio, and Boni, Benito, *The Concrete Architecture of Riccardo Morandi*, London: Alex Tiranti, 1965

Borrego, John, *Space Grid Structures*, Cambridge, MA: MIT Press, 1968

Burgess, S. C., and Pasini, D., 'Analysis of the structural efficiency of trees' in *Journal of Engineering Design*, Vol. 15, No. 2, April 2004, pp.177–193, Oxford: Taylor & Francis, 2004

Carter, Peter, *Mies van der Rohe at Work*, London: Phaidon, 1999

Chanakya, Arya, *Design of Structural Elements*, Oxford: Taylor & Francis, 2009

Chilton, John, *The Engineer's Contribution to Contemporary Architecture: Heinz Isler*, London: Thomas Telford, 2000

Cobb, Fiona, *Structural Engineer's Pocket Book*, Oxford: Butterworth-Heinemann, 2008

Coucke, P., Jacobs, G., Sas, P., and De Baerdemaeker, J., *Comparative Analysis of the Static and Dynamic Mechanical Eggshell Behaviour of a Chicken Egg*, Department of Agro-engineering and Economics, International Conference on Noise and Vibration Engineering, ISMA 23, September 16–18 1998, pp.1497–1502, Department of Mechanical Engineering, KU Leuven, Belgium, downloadable as a PDF from www.isma-isaac.be/publications/isma23

Coutts, M. P., and Grace, J., *Wind and Trees*, Cambridge: Cambridge University Press, 1995

Denny, Mark, *The Physical Properties of Spider's Silk and their Role in the design of Orb-webs*, Department of Zoology, Duke University, Durham, North Carolina, 1976, downloadable as a PDF from: jeb.biologists.org/content/65/2/483.full.pdf

Elliot, Cecil D., *Technics and Architecture*, Cambridge, MA: MIT Press, 1992

Engel, Heinrich; *Structure Systems*, New York: Van Nostrand Reinhold Company, 1981

Fisher, R. E., *Architectural Engineering – New Structures*, New York: McGraw-Hill, 1964

Fuller, R. B., *Inventions: The Patented Works of R. Buckminster Fuller*, New York: St. Martin's Press, 1983

Gole, R. S., and Kumar, P., *Spider's silk: Investigation of spinning process, web material and its properties*, Department of Biological Sciences and Bioengineering, Indian Institute of Technology Kanpur, Kanpur, 208016, downloadable as a PDF from: www.iitk.ac.in/bsbe/web%20 on%20asmi/spider.pdf

Goodchild, C. H., *Economic Concrete Frame Elements: A Pre-Scheme Design Handbook for the Rapid Sizing and Selection of Reinforced Concrete Frame Elements in Multi-Storey Buildings*, Surrey: British Cement Association, 1997

Gordon, J. E., *Structures: Or Why Things Don't Fall Down*, London: Penguin, 1978

Greco, Claudio, *Pier Luigi Nervi*, Lucerne: Quart Verlag, 2008

Heartney, E., *Kenneth Snelson: Forces Made Visible*, Stockbridge, MA: Hard Press Editions, 2009

Heyman, Jacques, *Structural Analysis: A Historical Approach*, Cambridge: Cambridge University Press, 1998

Hilson, Barry, *Basic Structural Behaviour*, London: Thomas Telford, 1993

Holgate, Alan, *The Work of Jörg Schlaich and his Team*, Stuttgart: Axel Menges, 1997

Hunt, Tony, *Tony Hunt's Structures Notebook*, Oxford: Architectural Press, 1997

Ioannides, S. A., and Ruddy, J. L., *Rules of Thumb for Steel Design* (paper presented at the North American Steel Conference), Chicago: Modern Steel Construction, February 2000, downloadable as a PDF from www.modernsteel.com/issue.php?date=February_2000

Kappraff, J., *Connections*, New York: McGraw-Hill, 1991

Khan, Y. S., *Engineering Architecture*, New York: Norton, 2004

Krausse, J., *Your Private Sky – Buckminster Fuller*, Zürich: Lars Müller Publishers, 2001

LeDuff, P., and Jahchan, N., *Eggshell Dome Discrepant Event*, Teacher's Guide SED 695B, 2005, http://www.csun.edu/~mk411573/discrepant/discrepant_event.html

Macdonald, A. J., *Structure & Architecture*, Oxford: Architectural Press, 2001

Margolis, I., *Architects + Engineers = Structure*, London: John Wiley & Sons, 2002

Mark, R., *Experiments in Gothic Structure*, Cambridge, MA: MIT Press, 1989

Megson, T. H. G., *Structural and Stress Analysis*, Oxford: Elsevier, 2005

Millais, M., *Building Structures*, London: E & F Spon, 1997

Morgan, J., and Cannell, M. G. R., *Structural analysis of tree trunks and branches: tapered cantilever beams subject to large deflections under complex loading*, Tree Physiology 3, pp.365–374, Victoria, BC: Heron Publishing, 1987, downloadable as a PDF from: treephys.oxfordjournals.org/content/3/4/365.full.pdf

Mosley, B., Bungey, J., and Hulse, R., *Reinforced Concrete Design*, Basingstoke: Palgrave, 2007

Nerdinger, W., *Frei Otto: Complete Works*, Basel: Birkhauser, 2005

Nerdinger, W. (et al), *Wendepunkte im Bauen Von der seriellen zur digitalen Architektur*, Munich: Edition Detail, 2010

Nervi, Pier Luigi, *Structures*, New York: F. W. Dodge Corporation, 1956

Nordenson, Guy (ed.), *Seven Structural Engineers: The Félix Candela Lectures*, New York: Museum of Modern Art, 2008

Otto, Frei, *Finding Form*, Fellbach: Edition Axel Menges, 1995

Popovic Larsen, O., *Reciprocal Frame Architecture*, Oxford: Architectural Press, 2008

Rice, P., *An Engineer Imagines*, London: Ellipsis, 1993

Sandaker, B., *The Structural Basis of Architecture*, Oxford: Routledge, 2011

Scott, Fred, *On Altering Architecture*, Oxford: Routledge, 2007

Steel Construction Institute, *Steel Designers' Manual*, Chichester: Wiley-Blackwell, 2005

Torroja, Eduardo, *Philosophy of Structures*, Los Angeles: University of California Press, 1958

Twentieth-Century Engineering, exhibition catalogue, New York: Museum of Modern Art, 1964

Veltkamp, M., *Free Form Structural Design: Schemes, Systems & Prototypes of Structures for Irregular Shaped Buildings*, Delft: Delft University Press, 2007

Wachsmann, K., *The Turning Point of Building*, New York: Reinhold, 1961

Wells, M., Engineers: *A History of Engineering and Structural Design*, Oxford: Routledge, 2010

Useful websites:
http://en.structurae.de
http://eng.archinform.net
http://designexplorer.net/
http://www.tatasteelconstruction.com

(websites accessed 10.10.12)

Index

Page numbers in italics refer to picture captions

Picture credits

Pictures by Will McLean, Pete Silver and Peter Evans, except where indicated below:
Cover images of Pompidou-Metz – photo: flashover/Alamy, drawings: designtoproduction, Zürich and Holzbau Amann
13 (1) David Scarf/Science Photo Library
13 (4) Carrier *USS Abraham Lincoln* (CVN72), U.S. Navy, photo by Photographer's Mate Airman Justin Blake
15 (2) Generated eggshell mesh using shell-type elements. (Based on a diagram from *Comparative Analysis of the Static and Dynamic Mechanical Eggshell Behaviour of a Chicken Egg* by P. Coucke, G. Jacobs, and J. De Baerdemaeker, Department of Agro-engineering and -economics, KU Leuven, Belgium, and P. Sas, Department of Mechanical Engineering, division PMA, KU Leuven, Belgium)
15 (3) ©Dennis Kunkel Microscopy, Inc./Visuals Unlimited/ Corbis Rights Managed
15 (4) ©Paul M.R. Maeyaert
15 (5) Courtesy MBM Arquitectes
17 (2) Laurence King Publishing
19 (1) ©Alexander Yakovlev/Fotolia
19 (2) ©Rick Rickman/NewSport/Corbis
20 (3) William Ruddock
20 (4) John Timbers/ArenaPAL
59 (1) ©[apply pictures]/Alamy
59 (2) ©travelbild.com/Alamy
59 (3) ©Visions of America, LLC/Alamy
59 (4) ©Tracey Whitefoot/Alamy
59 (5) ©Suzanne Bosman/Alamy
59 (6) ©Jon Bower Canada/Alamy
60 (1) ©Wiskerke/Alamy
60 (2) ©VIEW Pictures Ltd/Alamy
60 (3) ©J.D. Fisher/Alamy
60 (4) ©Michael Snell/Alamy
61 (1) ©imagebroker/Alamy
61 (2) ©VIEW Pictures Ltd/Alamy
61 (3) © Arcaid Images/Alamy
67 bottom right © PSL Images/Alamy
70 (1) iStockphoto/Thinkstock
70 (2) ©stockex/Alamy
70 (3) Stockbyte/Thinkstock
71, 72 top, 72 top centre iStockphoto/Thinkstock
72 bottom centre Comstock/Thinkstock
72 bottom Hemera/Thinkstock
80 top iStockphoto/Thinkstock
80 centre mambo6435/Shutterstock
80 centre bottom ©David R. Frazier Photolibrary, Inc./Alamy
80 bottom iStockphoto/Thinkstock
83 centre ©Tim Cuff/Alamy
89 (1) photo courtesy of www.nooksncorners.com
89 (2) Holger Knauf – www.holgerknauf.de
90 (4) Courtesy Atelier Frei Otto
91 (5) Courtesy Heinz Isler
92 RIBA (image no. 2845-23)
93 top ©Ian Cowe/Alamy
94 (3) Images courtesy of Dr. Arnold Wilson at the Brigham Young University Laboratories
95 (4) Images courtesy of MIT Masonry Research Group (MRG): John Ochsendorf, Mallory Taub, Philippe Block, Lara Davis, Florence Guiraud Doughty, Scott Ferebee, Emily Lo, Sze Ngai Ting, Robin Willis, Masoud Akbarzadeh, Michael Cohen, Samantha Cohen, Samuel Kronick and Fabiana Meacham
105 (1–4) Images courtesy of Prof. Robert Mark
108–9 Drawings by Akos Kovacs
114–115 Private collection, London
116 National Archives
117 (2) National Railway Museum/Science and Society Picture Library
117 (3) ©Toby/Fotolia
119 (2) Popperfoto/Getty Images
119 (3) ©Paul M.R. Maeyaert
120 scotlandsimages.com/Crown Copyright 2008. The National Archives of Scotland
121 ©Louise McGilviray/Fotolia
123 (1–4) Wikimedia Commons
123 (5–6) Photographs by Vladimir Schukov
125 (1) ©Bettmann/Corbis
125 (2) ©John Alexander Douglas Mucurdy/National Geographic Society/Corbis
129 (2) Courtesy: CSIC IETcc
129 (3) ©Bildarchiv Monheim GmbH/Alamy
133 (3) ©Cecil Handisyde-AA
133 (4 & 6) Luis M. Castañeda
133 (5) Jorge Ayala/www.ayarchitecture.com
135 (1–4 & 6–8) Courtesy of William Ruddock
137 (1) US Patent 3,197,927
138 (3–4) Images courtesy of © Karl Hartig
139 (5) Courtesy, The Estate of R. Buckminster Fuller
142–143 Images courtesy of Andrea Giodorno
151 (1) Daniel Schwen (Wikimedia Commons)
151 (3) Image courtesy of Axel Kilian, Designexplorer
153 (1) US Patent 4,059,937
157 (1) © Kenneth Snelson
163 (1–4) Images courtesy of Dante Bini, photography by Max Dupain
163 (5) Dante Bini
164 ©Trajano Paiva/Alamy
165 (3) ©Arcaid Images/Alamy
165 (4) ©MJ Photography/Alamy
167 Frank Oudeman
168–171 (1–10) Images courtesy of Dewhurst Macfarlane
172 Photograph by Richard Johnson, © Will Alsop, Alsop Architects, Archial Group
173 Photograph by Richard Johnson, © Will Alsop, Alsop Architects, Archial Group
174 © Will Alsop, Alsop Architects, Archial Group
175 (5–7) Images courtesy of expServices, Inc.
175 (8–12) © Will Alsop, Alsop Architects, Archial Group
179–81 Images courtesy of Bertus Mulder
182–85 Images courtesy of Ensamble Studio
188–89 Images courtesy of Junya Ishigami and Associates
191–93 Images courtesy of Price & Myers and M-Tec/WEC Group
193 (12) ©Nick Rearden
195 ©imagebroker/Alamy
196 (3, 4 & 6) Images courtesy of Holzbau Amann
196 (5) Image courtesy of designtoproduction, Zürich
197 (7) Image courtesy of designtoproduction, Zürich
197 (8–11) Images courtesy of Holzbau Amann
200–201 © Skidmore, Owings & Merrill LLP

Authors' acknowledgements

Jessica Brew
Philip Cooper
Liz Faber
Samantha Hardingham
Kate Heron
Eva Jiricna
Tim Macfarlane
Bert and Freda McLean
Robert Mark
Bertus Mulder
Christian Müller
Nils D. Olssen
William Ruddock
Esther Silver